Explorations
In Beginning and Intermediate Algebra
Using the TI-82

With Appendix Notes for the TI-85, Casio fx7700GE, and HP48G

Deborah J. Cochener
Bonnie M. Hodge

Austin Peay State University

D1501533

Brooks/Cole Publishing Company

I(T)P™   An International Thomson Publishing Company

Pacific Grove • Albany • Bonn • Boston • Cincinnati • Detroit • London • Madrid • Melbourne
Mexico City • New York • Paris • San Francisco • Singapore • Tokyo • Toronto • Washington

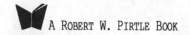A ROBERT W. PIRTLE BOOK

Sponsoring Editor: Elizabeth Barelli Rammel
Marketing Representative: Bob Schuh
Editorial Assistant: Linda Row
Production Editor: Tessa A. McGlasson
Cover Design: Vernon T. Boes
Printing and Binding: Patterson Printing

I(T)P The ITP logo is a trademark under license.

For more information, contact:

BROOKS/COLE PUBLISHING COMPANY
511 Forest Lodge Road
Pacific Grove, CA 93950
USA

International Thomson Editores
Campos Eliseos 385, Piso 7
Col. Polanco
11560 México D. F. México

International Thomson Publishing—Europe
Berkshire House 168-173
High Holborn
London WC1V 7AA
England

International Thomson Publishing GmbH
Königswinterer Strasse 418
53227 Bonn
Germany

Thomas Nelson Australia
102 Dodds Street
South Melbourne, 3205
Victoria, Australia

International Thomson Publishing—Asia
221 Henderson Road
#05–10 Henderson Building
Singapore 0315

Nelson Canada
1120 Birchmount Road
Scarborough, Ontario
Canada M1K 5G4

International Thomson Publishing—Japan
Hirakawacho Kyowa Building, 3F
2-2-1 Hirakawacho
Chiyoda-ku, Tokyo 102
Japan

Printed in the United States of America.

5 4 3 2 1

ISBN 0-534-34091-1

TEXAS INSTRUMENTS is a registered trademark owned by Texas Instruments.
HEWLETT-PACKARD is a registered trademark owned by Hewlett-Packard Company.
CASIO is a registered trademark owned by Casio Computer Company, Ltd.

DEDICATION

David and Sherah

Blaine, Blaire, and Trey

for their patience,
understanding,
support,
&
love

THANK YOU!

Tell me, and I'll forget.
Show me, and I may not remember.
Involve me, and I'll understand.

-Native American Saying

TABLE OF CONTENTS

Preface to the Instructor vii

Preface to the Student ix

Basic Calculator Operations

1. Getting Acquainted with Your TI-82 3
2. Focus on Fractions 17
3. Short Cut Keys: How to Be More Stroke Efficient 27
4. Order of Operations 33
5. Evaluating Expressions through Substitution 41
6. Introduction to Positive and Negative Exponents 49
7. Scientific Notation 59
8. Using the STOre Key to Evaluate Expressions and Check Solutions 65
9. Rational Exponents and Radicals 77

Graphically Solving Equations and Inequalities

10. Graphical Solutions to Linear Equations 89
11. Linear Applications 99
12. Graphical Solutions to Linear Inequalities 109
13. Graphical Solutions to Absolute Value Equations 119
14. Graphical Solutions to Absolute Value Inequalities 125
15. Solving Factorable Polynomial Equations 133
16. Solving Non-Factorable Quadratics 145
17. Applications of Quadratic Equations 151
18. Solving Quadratic Inequalities 161
19. Solving Nonlinear Inequalities 171
20. Solving Radical Equations 183

Graphing and Applications of Equations in Two Variables

21. How Does the TI-82 Actually Graph? (Exploring Points and Pixels) 193
22. Graphing Linear Equations in Two Unknowns 199
23. Preparing to Graph: Calculator Viewing Windows 209
24. Where Did the Graph Go? 219
25. Discovering Slope 223
26. The State Fair and Your Graphing Calculator 235
27. Applications of Systems of Linear Equations 239
28. Functions 245
29. Graphing Equations with Radicals 257
30. Discovering Parabolas 263
31. Exponential Functions and Their Inverses 269
32. Predict-A-Graph 281

Stat Plots

33. Statistics: Plotting Paired Data 287
34. Frequency Distributions 295
35. Line of Best Fit 305

Trouble Shooting 313

PREFACE TO THE INSTRUCTOR

This unique workbook/text provides the student the opportunity for guided exploration of topics in elementary and intermediate algebra using the TI-82 graphics calculator. Integration of technology into developmental studies algebra classes has grown significantly with the introduction of the graphing calculator. These calculators have brought relatively inexpensive technology into the students' hands. However, instructors are now faced with the problem of integrating the technology without sacrificing course content. These activities will enable the students to develop algorithms typically found in elementary and intermediate algebra courses, improving both their understanding and retention of the material.

The text is written in a style that is both teacher and student friendly. It is designed to be used as a supplement to an existing text. The units provide springboards for further investigation of mathematics, either independently or in groups.

Having been used by two or more instructors over a period of two years, the units have enabled our students to become eager participants rather than passive observers of mathematics. The workbook was written with the belief that students who are active contributors in the classroom increase their own understanding and their long term retention of material. Of primary importance, however, is the fact that the units have provided the means and the opportunity for students to create their own mathematics.

Features

*Each unit provides guided exploration of a topic. Units are not meant to be done in numerical order, but rather according to concept.

*The workbook may be used over a period of more than one semester/quarter as the student progresses through his/her mathematics sequence.

*As an institution changes textbooks, this supplement need not be changed.

*Cross-referencing between units as well as a correlation chart relating course topics to workbook units is provided.

*A key correlation chart that shows in which units keys are introduced is also provided.

*Answers are provided either within the text of the unit or at its end.

*Applications are provided both within individual units and as separate units.

*Space and topical outlines for student summaries are provided at the end of each unit.

*A Troubleshooting Section is provided at the end of the text. It contains common student errors as well as explanations of the error screens students most often encounter.

*Units are written in a manner that enables the student with little or no algebraic experience to read and explore independently.

*The units provide springboards for both classroom discussion and further investigation either individually or as a class.

Organization

The workbook is divided into four sections. Each section contains both a concept correlation chart and a key introduction chart.

UNITS #1-9: These units introduce the student to the calculator and its capabilities in performing computational tasks.

UNITS #10-20: These units provide the link between the "by hand" processes traditionally taught in elementary and intermediate algebra courses and the "picture" associated with the algebraic expressions and equations. Students and instructors are encouraged to work through these units BEFORE the student has been formally introduced to graphing.

UNITS #21-32: These units introduce the student to graphing as a formal process. It is in these units that adjusting view screens (WINDOWS) and interpreting graphs is emphasized.

UNITS #33-35: These units cover basic statistics. They are not meant to be a comprehensive coverage of the subject, but rather an introduction to the capabilities of the TI-82 in basic data analysis.

Preface for the Appendices

Although we recommend that the student use a TI-82 graphing calculator, we realize that may not always be possible. As a result, we have written appendices for students using either the TI-85, Casio fx-7700GE, or HP 48G. When appropriate, we have included an appendix for each calculator at the end of the unit.

The appendix will point out the appropriate keying sequence in performing the operations that are required for a particular unit. We do not include keystrokes for every example or for every occurrence of a TI-82 command. The keystrokes necessary for performing the examples and exercises are usually presented only once. In some cases, there are no analogous operations between the different calculators.

We strongly recommend that students who are not using a TI-82 write down a summary sheet of TI-82 commands and their calculator's equivalents. This is useful in remembering commands as students work through the units.

In using the appendices, keep in mind that we have attempted to point out the significant differences between the calculators. Therefore, the student should not ignore the instructions for the TI-82. Also, the student should read the manual accompanying their calculator to familiarize themselves with the basic operations of their calculator.

Since many of the exercises are geared specifically towards the TI-82, students should work with other classmates using the TI-82 in order to answer the questions and understand how to use the TI-82.

Summary

This workbook is unique in that it teaches students to use the available technology AND to link that technology to the mathematics. It does not cover all of the capabilities of the TI-82, but provides information that works well as a springboard for further student investigation. The primary value, however, is that the workbook provides students with problem-solving approaches, helps them construct problem-solving strategies, and promotes critical thinking skills. (NOTE: Section specific correlation charts for any elementary and/or intermediate algebra text are available upon request from Brooks Cole Publishing.)

PREFACE TO THE STUDENT

This workbook was developed with YOU in mind. It is written in a style that is easily understood by students at varying levels of mathematical proficiency. The following are suggestions for using the workbook:

*READ and follow the instructions *slowly* and *carefully*. Pay close attention to detail.

*ALWAYS complete the summaries at the end of the units. This is where you bring your thoughts together and put the mathematics into language that is meaningful to YOU.

*Keep a log of those calculator techniques that have proven helpful.

*In this log, also write any questions and/or comments that occur to you as you work through the units. Form a study group with your classmates to discuss these entries. WHEN YOU DISCUSS MATHEMATICS, YOU LEARN MATHEMATICS.

*The calculator is a TOOL for learning and doing mathematics. If an answer does not seem reasonable, always double check your reasoning, keystrokes and screen.

*The units are correlated to algebraic concepts - be sure you work through prerequisite units.

*Keep this workbook; it will serve as a reference for future courses - one that you have personalized.

We hope that by learning to use the TI-82 calculator your confidence in your abilities to create and DO mathematics increases. Experiment frequently with the keys and menu options that are not covered in the workbook. Last, but not least, HAVE FUN!

Preface for the Appendices

Although we recommend that the student use a TI-82 graphing calculator, we realize that may not always be possible. As a result, we have written appendices for students using either the TI-85, Casio fx-7700GE, or HP 48G. When appropriate, we have included an appendix for each calculator at the end of the unit.

The appendix will point out the appropriate keying sequence in performing the operations that are required for a particular unit. We do not include keystrokes for every example or for every occurrence of a TI-82 command. The keystrokes necessary for performing the examples and exercises are usually presented only once. In some cases, there are no analogous operations between the different calculators.

We strongly recommend that students who are not using a TI-82 write down a summary sheet of TI-82 commands and their calculator's equivalents. This is useful in remembering commands as students work through the units.

In using the appendices, keep in mind that we have attempted to point out the significant differences between the calculators. Therefore, the student should not ignore the instructions for the TI-82. Also, the student should read the manual accompanying their calculator to familiarize themselves with the basic operations of their calculator.

Since many of the exercises are geared specifically towards the TI-82, students should work with other classmates using the TI-82 in order to answer the questions and understand how to use the TI-82.

ACKNOWLEDGMENTS

We are grateful and appreciative for the input of time and expertise by those who reviewed this workbook at the various stages of its development:

Judy Darke, Richview Middle School
Mary Lou Hammond, Spokane Community College
Don Shriner, Frostburg State University
Patricia Stone, Tomball College

We appreciate the guidance and support of Elizabeth Rammel and Bob Pirtle, our editors at Brooks/Cole publishing. Elizabeth has always had the right answer to the question and has made the writing of this workbook a pleasure. Her support has been unwavering, and her belief in our abilities steadfast. We thank George Seki and his staff at Laurel Technical Services for their appendices material and Patricia Stone for her contributions to the "DOs and DON'Ts" list. Dave Gustafson has earned our gratitude and thanks - if not for his belief in our abilities this workbook would never have become a reality. Finally, we would like to again thank our husbands, David and Blaine, and our children, Sherah, Blaire, and Trey, for their support and love. They have always believed in us and it is they who made the dream become reality. Our families' contributions to the editing, reviewing, wording, and the testing of material has been invaluable.

CORRELATION CHART

UNIT * Instructors are encouraged to use Units marked by an * to introduce concepts.	PRE-REQ. UNIT	CORRELATING CONCEPT
#1 Getting Acquainted with your TI-82		Signed numbers, absolute value, square roots, and exponents
#2 Focus on Fractions	#1	Fraction arithmetic
#3 Short Cut Keys: How to be More Stroke Efficient	#1, 2	A tour of the short cut keys.
#4 Order of Operations *	#1, 2	This unit can be used to develop the Orders of Operations Rules
#5 Evaluating Expressions through Substitution	#1, 2, 4	Substitution Property
#6 Introduction to Positive and Negative Exponents *	#1 - 3	Examines the effect of negative exponents and the zero exponent
#7 Scientific Notation *	#1 - 3	Scientific Notation (This unit may be used to replace in-class instruction.)
#8 Using the STOre Key to Evaluate Expressions and Check Solutions	#1 - 4	Use of the Substitution Property to validate polynomial operations and check solutions to equations.
#9 Rational Exponents & Radicals *	#1 - 3, 8	This unit can be used to introduce the relationship between rational exponents and radicals.

INTRODUCTION OF KEYS

Unit Title	Keys
#1: Getting Acquainted With Your TI-82	ON/OFF 2nd ▲ ▼ (to darken/lighten) MODE ENTER CLEAR (-) Abs ( ) parentheses √ x^2 ∧
#2: Focus on Fractions	MATH 1:▸Frac
#3: Short Cut Keys: How to be More Stroke Efficient	QUIT INS DEL π ENTRY ANS
#4: Order of Operations	No new keys
#5: Evaluating Expressions Though Substitution	No new keys
#6: Introduction to Positive and Negative Exponents	No new keys
#7: Scientific Notation	MODE (SCI)
#8: Using the STOre Key to Evaluate Expressions and Check Solutions	STO▸ X,T,θ ALPHA :
#9: Rational Exponents and Radicals	MATH 4: $\sqrt[3]{}$ 5: $\sqrt[x]{}$

UNIT #1
GETTING ACQUAINTED WITH YOUR TI-82

Touring the TI-82

Take a few minutes to study the TI-82 graphing calculator. The keys are color-coded and positioned in a way that is user friendly. Notice there are dark blue, black, and gray keys, along with a single light blue key.

<u>Dark blue keys:</u> On the right side of the calculator are the dark blue keys. At the top are 4 directional cursor keys. These may be used to move the cursor on the screen in the direction of the arrow printed on the key. The 4-operation symbols (addition, subtraction, multiplication, and division) are also in dark blue. Notice the key marked "ENTER". This will be used to activate commands that have been entered, thus there is no key on the face of the calculator with the equal sign printed on it.

<u>Gray keys:</u> The 12 gray keys that are clustered at the bottom are used to enter digits, a decimal point, or a negative sign. Notice the gray key beneath the light blue key in the upper left position, labeled "ALPHA". We will come back to this key shortly; since it's position is different from the other gray keys it serves a different function.

<u>Black keys:</u> The majority of the keys on the calculator are black. Notice the [X,T,θ] key in the second row and second column and the 5 that are positioned beneath the screen. These are used for graphing; discussion of them will begin in Unit 10. The "ON" key is the black one located in the bottom left position.

<u>Light blue key:</u> The only key that is light blue is the one in the upper-left position labeled "2nd".

Above most of the keys are words and/or symbols printed in either light blue or white. To access a symbol in light blue (printed above any of the keys) you must first press the light blue [2nd] key, and then the key BELOW the symbol (function) you wish to access. For example, to turn the calculator "OFF" we notice that the word "OFF" is printed in light blue above the "ON" key. Therefore, we must press the keys [2nd] [ON] to turn the calculator off. You do these keystrokes *sequentially* - not simultaneously. Throughout this book the following symbolism will be used. Actual keys will be denoted in brackets, [ ], whereas those symbols written above a key will be denoted with < >. Thus, the previous command for turning the calculator off would appear as [2nd] <OFF>. The symbols [ ] or < > will cue you <u>where</u> to look for a command - either printed on a key or above it.

To access a letter of the alphabet (printed in white above some of the keys), or any other symbol printed in white above a key, first press the gray "ALPHA" key and then the key *below* the desired letter or symbol. Again, the keystrokes are sequential.

Below the screen are 5 black keys labeled **[Y=]**, **[WINDOW]**, **[ZOOM]**, **[TRACE]**, and **[GRAPH]**. These keys are positioned together below the screen because they are used for graphing functions.

Note: The TI-82 has an **A**utomatic **P**ower **D**own (**APD**) feature which turns the calculator off when no keys have been struck for several minutes. When this happens, press **[ON]** and you will return to the screen that you were working on.

Let's Get Started!

Turn the calculator on by pressing **[ON]**. If you are having trouble seeing the display, press **[2nd] [▲]** to darken the screen, or **[2nd] [▼]** to lighten the screen. Notice that when you press **[2nd]** an arrow pointing up appears on the blinking cursor.

To ensure that we are in the desired mode, press **[MODE]**. All of the options on the far left should be highlighted. If not, use the **[▼]** to place the blinking cursor on the appropriate entry and press **[ENTER]**. Exit MODE by pressing **[CLEAR]**. You should now be at the home screen. This is where you will enter your expressions. Press **[CLEAR]** until the screen is cleared except for the blinking cursor in the top left corner.

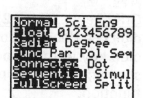

Integer Operations

Simplify the following expressions on the TI-82. Answers have been provided so that you may ensure correct entry. **RECORD YOUR CALCULATOR SCREEN LINE BY LINE EXACTLY** as it appears. Use the space provided below each problem.

If you get an error message refer to the "Trouble Shooting" section at the end of the text.

Example: Simplify: -8 - 2

> **Keystrokes: [(-)] [8] [-] [2] [ENTER]**
> **Screen display:**

Notice that you must differentiate between the operation of subtraction (the dark blue key on the right) and the negative sign (the gray key at the lower right).

Simplify each expression below. Record your calculator screen display in the space provided.

1. -6 + (-2)   ANS. -8

2. -14(-2)(-3)(2)   ANS. -168

3. 10 ÷ 2 + 7 - 3   ANS. 9

4. 18 - ˉ3   ANS. 21

5. -8 - 3   ANS. -11

Absolute Value

Absolute value is located above the [x⁻¹] key. To access it, press [2nd] <abs>.

Example: Simplify: |-3-2|

NOTE: We would read this as "the absolute value of the quantity negative three minus two". Keep this in mind as you enter the expression.

Keystrokes: [2nd] <abs> [(] [(-)] [3] [-] [2] [)] [ENTER]

Screen display:

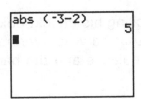

Simplify each of the following. Record your calculator screen in the space provided.

6. |-8| ANS. 8

7. |8| ANS. 8

8. $-|-2|$ ANS. -2

9. $|3-4|$ ANS. 1

10. $-|-2-4|$ ANS. -6

11. (a) Draw a number line (labeling integers), and find the distance between the points with coordinates -4 and 3 by counting units.

Distance is always positive - if direction is necessary, it is included through the use of words (such as North, South, etc.) or sign (positive or negative). To ensure that distances are always positive when doing computation, we find the distance between two points on a number line by using the concept of absolute value.

(b) The distance between two points can be found by computing *the absolute value of the difference between the coordinates.* Find the distance between the above two coordinates using this concept. Record your screen display.

12. Applications: While reconciling his checkbook, Trey recorded a balance of $31.00. The bank stated that he was overdrawn $22.00. What is the difference between Trey's balance and the bank's balance?   ANS. $53

13. Explain why parentheses are necessary in #9 and #10. (Hint: you may want to enter the expression in #9 without parentheses and compare your answer to the correct answer.)

Square Roots

Square root is located above the [x²] key. To access it, press **[2nd]** < √ >.

Example: Simplify: $\sqrt{25}$

Keystrokes: **[2nd]** < √ > **[2] [5] [ENTER]**

Screen display:

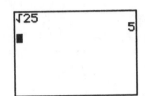

Simplify each expression below. Record the calculator display in the space provided.

14. $\sqrt{36}$ ANS. 6

15. $-\sqrt{49}$ ANS. -7

16. $\sqrt{25-16}$ ANS. 3

17. $\sqrt{72}$ ANS. 8.485281374

18. $\sqrt{36-8}$ ANS. 5.291502622

19. Explain why parentheses are necessary in #16 and #18. Again, you may want to enter the expression in two ways - once with and once without parentheses. Compare your results.

Powers of Numbers

You may square a number by either pressing the $[x^2]$ key after entering your number, or by using the caret [^] and then entering the desired exponent.

Example: Simplify: 4^2

Keystrokes: [4] $[x^2]$ [ENTER]   **OR**   **Keystrokes:** [4] [^] [2] [ENTER]

Screen display: **Screen display:**

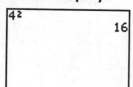   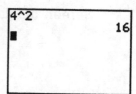

To raise to the third power (or higher), use the caret key.

Example: Simplify: $4(3)^5$

Keystrokes: [4] [(] [3] [)] [^] [5]

Screen Display:

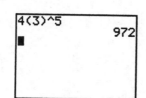

Simplify each of the following in two ways: "by hand" and "by calculator". Show each step of your work ("by hand") and record the display screen ("by calculator").

20. 3^4 ANS. 81

21. -3^4 ANS. -81

22. $(-3)^4$ ANS. 81

23. $-(3)^4$ ANS. -81

24. (-3^4) ANS. -81

Simplify each of the following expressions. Record your screen display in the space provided.

25. $\dfrac{3^3}{9}$

ANS. 3

26. $|4^5 - {}^-6^2|$

ANS. 1060

27. $\sqrt{5^3 - 10^2}$

ANS. 5

28. $(6^2 - 4^2)^3$

ANS. 8000

29. $(15 - 2)^3$

ANS. 2197

30. Troubleshooting: Each of the problems below has been entered **incorrectly** on the calculator. Make the necessary corrections so that the calculator display accurately represents the problem given. Be sure to verify with your calculator!

a. -4 - 3

```
Ans-4-3
```

b. $\sqrt{9 + 16}$

```
√9+16
```

9

c. $|-4 - (-11)|$

```
abs -4-11
```

31. Summarizing Results: Write a summary of what you have learned in this unit.
     Do not focus on keystrokes, but rather on the "big ideas" you have discovered
     while working the problems throughout the unit. Your summary should
     include the following points:
     a. use of the gray [(-)] key
     b. use of parentheses in entering expressions for absolute value, roots and
         exponents.

UNIT #1 APPENDIX
TI-85

Touring the TI-85

There are several differences between the TI-82 and the TI-85. The first is the keyboard. The arrow keys on the TI-85 are gray, and there are no dark blue keys. The [2nd] key is yellow-orange and is used to access the functions printed in yellow above the keys. The [ALPHA] key is light blue, and is used to access the alphabet and other symbols printed in light blue above the keys.

The TI-85 also has a set of five keys at the top of the keyboard which are called the menu keys. The current functions of menu keys are shown at the bottom of the display. In this text we will follow a menu key by its current function in parentheses.

In addition, many of the actual functions printed on the TI-85 keyboard are different from those printed on the TI-82 keyboard. Whenever possible, the equivalent instructions for the TI-85 will be listed in the appendix to each section. For instance, under <u>Let's Get Started!</u> you are instructed to press [MODE] to access the mode screen. The equivalent instruction for the TI-85 is to press [2nd] <MODE> to see the screen at right.

```
Normal Sci Eng
Float 012345678901
Radian Degree
RectC PolarC
Func Pol Param DifEq
Dec Bin Oct Hex
RectV CylV SphereV
dxDer1 dxNDer
```

Absolute Value

You will notice that there is no [x⁻¹] key nor an <ABS> on the TI-85. However, one can be *created* using the [CUSTOM] key. Using this key, you can access up to 15 frequently used functions with just two keystrokes. But first you must *choose* the functions to put on the custom menu. To do this, first select [2nd] <CATALOG> and push [F3] (CUSTM). Place the arrow next to "abs" in the list, then push [F1]. You will see "abs" listed under "PAGE↓". Do the same procedure to add other functions to the [CUSTOM] menu as you need them, each time selecting an open spot in the menu by choosing the menu key below the open spot. When you are done, press [EXIT] until the menus at the bottom of the screen clear, to return to the main window.

Example: Simplify | -3 - 2 |
    (Ensure you are at the main window - press [EXIT] if necessary.)
    Keystrokes: **Screen display:**
    [CUSTOM] [F1] [(] [(-)] [3] [-]
    [2] [)] [ENTER]

```
abs (-3-2)
                        5



abs  ▶Frac
```

```
┌─────────────────────────────────────────────────┐
│                                                   │
│              UNIT #1  APPENDIX                    │
│              Casio fx7700GE                        │
│                                                   │
└─────────────────────────────────────────────────┘
```

Touring the Casio fx-7700GE

The TI-82 and the Casio *fx7700GE* are substantially different machines. Many of the capabilities of the TI-82 are comparable to the Casio, but they are accessed in different ways. Further, the screen outputs are quite dissimilar. We recommend if you have a Casio that you work through the Quick-Start section in its manual to become familiar with your calculator.

The goal of this appendix will be to explain to you Casio's syntax in performing the tasks in this manual. However a short introduction to the Casio *fx-7700GE*, hereafter referred to as the Casio, is in order.

The key feature of the Casio is the MAIN MENU. To access this, turn it on by pressing [AC/ON] and then [MENU]. This screen allows you to select from among the several features of the calculator. For example, to adjust the darkness of the screen, use the arrows to move down to **CONT** and press [EXE]. Now you can use the left and right arrows to adjust the contrast of the screen. To return to the MAIN MENU, press [MENU].

The keyboard is similar to the TI-82. There are several gray keys, which are the basic numbers and operations; several black keys, which give most commonly used mathematical functions; six green function keys below the screen, which are used to access the menu options; and three colored keys. The orange [SHIFT] key is used to access all the words printed in orange above the other keys. The red [ALPHA] key is used to access the letters of the alphabet and other symbols which are printed in red above the other keys. Finally, the blue [EXE] key is used to execute many functions on the calculator.

Absolute Value

The absolute value function, which is used just as it would be on the TI-82, is found by first entering the **COMP** mode from the MAIN MENU. Press [SHIFT] <MATH> (located above [5]). Then select [F3] for (NUM) and choose the first option, [Abs], by pressing [F1].

Example: Simplify: |-3 - 2|

Keystrokes: [SHIFT] <MATH> [F3] [F1]
    [(] [(-)] [3] [-] [2] [)] [EXE]

Screen display:
```
┌──────────────────────┐
│Abs (-3-2)            │
│                   5. │
│                      │
│                      │
│                      │
│Abs│Int│Frc│Rnd│Intg │
└──────────────────────┘
```

UNIT #1 APPENDIX
HP 48G

Touring the HP 48G

Read the first chapter of the User's Guide to familiarize yourself with the keyboard and display of the HP 48G.

The keyboard has six levels of functions. The primary keyboard is labeled on the buttons. The left-shift keyboard is labeled in purple above and to the left of the primary buttons. This keyboard is activated when by pressing the purple key located on the left side of the key pad. We will use [←] to denote this key. (Note that we will use [⇐] to denote the key in the middle of the right side of the key pad.) The right-shift keyboard is labeled in green above and to the right of the buttons. This keyboard is activated by pressing the green key located on the left side of the key pad and below [←]. We will use [→] to denote this key. The Alpha keyboard is activated by pressing [α] which is located above [←]. The Alpha keys are all the capital letters. These keys are labeled in white to the right of the buttons. The Alpha left-shift keyboard is activated by pressing [α] [←]. The alpha left-shift keys include the lowercase letters, along with some special characters. These keys are not labeled on the calculator. The Alpha right-shift keyboard is activated by pressing [α] [→]. The Alpha right-shift keys include Greek letters, along with some special characters. These keys are also not labeled on the calculator. The Alpha keyboards are displayed in the User's Guide.

At the top of the calculator, there are six blank menu keys. The current commands associated with the menu keys are shown at the bottom of the display screen. We will symbolize menu keys by parentheses, ( ), with the function name inside the parentheses.

Let's Get Started

andTo adjust the contrast on the HP 48G, hold down the ON button   press [+] to darken or [-] to lighten.

To ensure that we are in the desired mode, press [→] <MODES>. The options should be set as follows:

If not, use the arrow keys to move onto an option. Press (CHOOS) to change the option. Use the arrow keys to select the appropriate option and press [ENTER] or (OK).

The Stack

The HP 48G comes with a Quick Start Guide. You should look at the first few chapters to learn how to use its basic operations.

The HP 48G is significantly different from both the TI and Casio calculators. One difference is the use of the stack for the storage of numbers and other objects. The storage locations are called levels 1, 2, 3, etc. The number of levels depends on the number of objects stored. The calculator will always display the first few levels. As new data is entered onto the stack, the new data moves to level 1 and the old data moves up one level.

Commands require data to be present before being executed. The data used is removed from the stack upon execution of the command. The result from the command is entered onto the stack.

Algebraic-Entry Mode

If you are not accustomed to using the stack, you can enter an expression in manner similar to the TI's and Casio by pressing ['], then entering the expression.
To evaluate the expression press **[EVAL]**.

EquationWriter

The EquationWriter is an application that can be used to enter algebraic expressions and equations in a form close to the way you would using paper and pencil. To activate the EquationWriter press **[←] <EQUATION>**. Then enter your expression as you would on a TI or Casio. When you are done, press **[ENTER]**. To evaluate the entry expression, press **[EVAL]**.

Integer Operations

Example: Simplify: - 8 - 2

> Keystrokes (Stack): **[8] [+/-] [ENTER] [2] [-]**
> Keystrokes (Algebraic): **['] [+/-] [8] [-] [2] [ENTER] [EVAL]**
> Keystrokes (EquationWriter): **[←][<EQUATION> [+/-] [8] [-] [2]**
> **[ENTER] [EVAL]**

Absolute Value

The absolute value function is a MENU command and is found in the MTH REAL menu. To access the absolute value function press **[MTH] (REAL) [NXT]**. ABS will appear on the menu display corresponding to the first menu key. Note that you press **[NXT]** to display more commands on the menu. ABS will remain on the menu display until the menu is changed.

Example: Simplify: | - 3 - 2 |

       Keystrokes (Stack): [3] [+/-] [ENTER] [2] [-] (ABS)
       Keystrokes (Algebraic): ['] (ABS) [+/-] [3] [-] [2] [ENTER] [EVAL]
       Keystrokes (EquationWriter): [←][<EQUATION> (ABS) [+/-] [3] [-] [2]
           [ENTER] [EVAL]

Square Roots

You may take the square root of a number by pressing [$\sqrt{x}$] after you enter your

number into the stack. If you are using algebraic entry, press ['] [$\sqrt{x}$] then enter
your number, **[ENTER] [EVAL]**.

Powers of Numbers

The square key is the left shift of the square root key. To raise a number by an
exponent, enter the number into the stack, then enter the exponent and press [y^x].

Example: Simplify: 4(3)^5

       Keystrokes (Stack): [4] [ENTER] [3] [ENTER] [5] [y^x] [×]
       Keystrokes (Algebraic): ['] [4] [×] [←] <( )>[3] [▶] [y^x] [5] [ENTER] [EVAL]
       Keystrokes (EquationWriter): [←][<EQUATION> (ABS) [4] [×] [←] <( )>[3]
           [▶] [y^x] [5] [ENTER] [EVAL]

The TI-82 can be used to perform arithmetic operations with fractions. Pressing the [MATH] key reveals a math menu. Take a minute to use the [▼] cursor to scroll down the menu. Notice that there are 10 options available. The first option, 1:▶ Frac, is the one that we will now investigate.

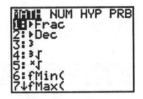

1.   Enter the fraction 1/2 by pressing [1] [÷] [2] [MATH]. Under the MATH menu "1:▶Frac" is highlighted. Because it is highlighted we can now press [ENTER] to select "1". Press [ENTER] again to activate the "convert to a fraction" command. Notice that the calculator simply displays our fraction.

2.   Re-enter 1/2 by using the following keystrokes: [1] [÷] [2] [ENTER]. Notice that this time we did not access the MATH menu and the "▶Frac" command. The TI-82 converts our expression to a decimal (as will any scientific calculator). However, if we want a fraction rather than a decimal, we can now press [MATH] [ENTER] (again, the ▶Frac option is chosen because it is highlighted). The calculator now displays **Ans** ▶**Frac** and our decimal is converted back to the fractional form once [ENTER] is pressed.

3.   Enter the fraction 12/24 and access the ▶FRAC option. How does the calculator display this fraction?

      Enter the fraction 17/51 and access the ▶FRAC option. Again, how does the calculator express this fraction?

      What can you conclude about the way the TI-82 will *always* display fractional answers?

4. Read the mixed number "$3\frac{1}{2}$" aloud. Did you say the word "and"? This is our clue as to the way we enter mixed numbers on the TI-82. Enter $3\frac{1}{2}$ by pressing [3] [+] [1] [÷] [2] [MATH] [ENTER] [ENTER]. Record the calculator display in the space below.

Addition and Subtraction of Fractions

Simplify each expression below. Express all answers as fractions.
Record your screen display in the space provided.

Example: $\frac{1}{2} + \frac{1}{3}$

   Keystrokes: [1] [÷] [2] [+] [1] [÷] [3] [MATH] [ENTER] [ENTER]

   Screen display:

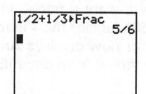

```
1/2+1/3▶Frac
          5/6
■
```

1. $\frac{2}{3} + \frac{3}{4} - \frac{1}{2}$ ANS. $\frac{11}{12}$

2. $\frac{5}{8} - \frac{9}{10}$ ANS. $-\frac{11}{40}$

3. $\frac{5}{8} + \frac{6}{7}$ ANS. $\frac{83}{56}$

4. $2\frac{1}{2} + 3\frac{2}{3}$ ANS. $\frac{37}{6}$

5. $\dfrac{6}{5} - \dfrac{1}{5}$ ANS. 1

Multiplication and Division of Fractions

Example: $\dfrac{2}{3} \div \dfrac{1}{5}$

     Keystrokes: [(] [2] [÷] [3] [)] [÷] [(] [1] [÷] [5] [)] [MATH] [ENTER] [ENTER]

     Screen display:

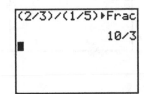

Simplify the following expressions, recording your screen display in the space provided.

6. $\dfrac{6}{5} \cdot \dfrac{1}{2}$ ANS. $\dfrac{3}{5}$

7. $\dfrac{4}{3} \div \dfrac{2}{5}$ ANS. $\dfrac{10}{3}$

8. $\dfrac{2}{3} \div \dfrac{2}{9}$ ANS. 3

9. $\left(-\dfrac{1}{2}\right)\left(\dfrac{3}{4}\right)$ ANS. $-\dfrac{3}{8}$

10. $\left(-\dfrac{2}{3}\right) \div \left(\dfrac{5}{8}\right)$ ANS. $-\dfrac{16}{15}$

Mixed Operations with Fractions

Simplify each expression below. Record your screen display in the space provided.

Example: $\left(\dfrac{2}{3}\right)^2 + \dfrac{1}{5}$

Keystrokes: [(] [2] [÷] [3] [)] [∧] [2] [+] [1] [÷] [5] [MATH] [ENTER] [ENTER]

Screen display:

```
(2/3)^2+1/5▶Frac
              29/45
■
```

11. $\left|-\dfrac{3}{5}\right| \cdot \sqrt{\dfrac{1}{4}}$

ANS. $\dfrac{3}{10}$

12. $\left(\dfrac{1}{2}\right)^3 \cdot \left(\dfrac{2}{3}\right)$

ANS. $\dfrac{1}{12}$

13. $\sqrt{\dfrac{4}{25}} \div 2$

ANS. $\dfrac{1}{5}$

14. $\left(\dfrac{3}{5}\right)^2 \cdot \left(\dfrac{2}{3}\right)^3$

ANS. $\dfrac{8}{75}$

15. $\dfrac{3}{5} \div \dfrac{1}{2} - \dfrac{5}{8}$

ANS. $\dfrac{23}{40}$

For each problem below, write an appropriate algebraic expression. Enter your expression on the calculator and record your screen display.

16. David makes $33,000 a year as a computer analyst. Approximately one-fourth of his salary is taken out for taxes.

    (a) How much of his salary goes to taxes?   ANS. $8250

    (b) How much of his salary does he take home?   ANS. $24750

17. The formula for the area of a triangle is $A = \dfrac{1}{2}bh$, where b is the base of the triangle and h is the altitude. Find the area of the pictured triangle.

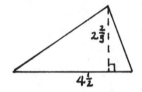

                                             ANS. 6 sq.units

8. Troubleshooting: Each of the problems below has been entered **incorrectly** on the calculator. Make the necessary corrections so that the calculator display accurately represents the problem given. Be sure to verify with your calculator!

    a. $\sqrt{\dfrac{25}{81}}$

    b. $\left| \dfrac{1}{4} + -\dfrac{3}{5} \right|$

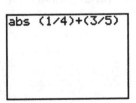

19. **Summarizing Results:** Summarize what you learned in this unit about entering fractions on the calculator. Go back through the unit and look at the displays you copied for problems as well as written responses to questions. You should include in your summary the use of parentheses when entering fractions on the TI-82.

20. Highlight #1 and #2 for your QUICK REFERENCE list of operations.

Solutions: The correct keystrokes to #18 are:

a. [2nd] <√> [(] [2] [5] [÷] [8] [1] [)]
b. `[2nd] <abs> [(] [1] [÷] [4] [+] [(-)] [3] [÷] [5] [)]

```
┌─────────────────────────────────────────────┐
│            UNIT #2  APPENDIX                  │
│                 TI-85                         │
└─────────────────────────────────────────────┘
```

The <MATH> key is located above the multiplication key on the TI-85. Press [2nd]
<MATH> and five menus will be displayed. Press [F5] (MISC). The ► at the right
end of the menu display indicates that there is more to view; press [MORE]. You
should see the screen displayed at the right. At this point, there
are two rows of menus displayed on your screen. The F1-F5 keys
access the lower row of menus while the M1-M5 keys access the
upper row. (To activate M1-M5 you must first press [2nd].)

Since we will be using the ►Frac function frequently, you will want to add it to your
CUSTOM menu. To do this, select [2nd] <CATALOG> and push [F3] (CUSTM).
Place the arrow next to "►Frac" in the list (it is found near the *bottom* of the list).
The quickest way to find "►Frac" is to press [▲] then [F2] (PAGE↑) (or [2nd] <M2>
(PAGE↑) if you have activated the CUSTOM menu). Then push [F2] or another open
F key. If none are open, press [MORE] and find an empty spot. If you must delete a
function from the [CUSTOM] menu to make room, you may do so by selecting [2nd]
<CATALOG> and [F4] (BLANK) (or [2nd] <M4>(BLANK) if you have activated the
CUSTOM menu), then choose a function to delete by selecting the appropriate menu
key. The ►Frac function can now be accessed from the CUSTOM menu.

NOTE: In Unit #1 we entered absolute value on our CUSTOM menu. If you press
[2nd] <MATH> [F1] (NUM) you will see that absolute value is actually located
under the MATH menu.

There is no "▸Frac" command on the Casio. To express a number as a fraction, you use the [a b/c] key on the left edge of your calculator keypad. For example, to enter 1/2 as a fraction, press [1] [a b/c] [2] [EXE]. The screen will display:

If you want to view 1/2 as a decimal, press the [a b/c] key. Press the [a b/c] key again to view as a fractional number. Note that you can only convert numbers that were previously displayed as fractional numbers from decimal numbers to fractional numbers. To enter the mixed number $3\frac{1}{2}$, press [3] [a b/c] [1] [a b/c] [2] [EXE]. (See display.)

Consult the Casio manual for more explicit instructions.

Mixed Operations with Fractions

One minor drawback is that the Casio will convert fractions to decimals whenever a fraction is raised to a power using the [^] button. Fractions will remain fractions if the [x^2] or <x^{-1}> keys are pressed.

To perform arithmetic operation with fractions either use the algebraic-entry mode or the EquationWriter.

To enter a fraction using the EquationWriter, press [←] [EQUATION] to enter the EquationWriter. Press [▲] to begin a numerator, after entering the numerator press [▼] (or [►]) to move from the numerator to the denominator. Press [ENTER] when you are done to exit the equation editor.

If you are using the algebraic-entry mode, press ['] then enter the numerator. Press [÷] then enter the denominator. Press [ENTER] when you are done.

You can then perform algebraic operations with fractions entered into the stack. Or you can enter the entire expression using the algebraic-entry mode or the equation editor. You must press [EVAL] to evaluate the expression. The result will be given in decimal form.

In order to convert decimal numbers to fractions, press [←] <SYMBOLIC> [NXT] to get →Q on the menu display. It will be the third item from the left, so it will correspond to the third menu key. As in the case of ABS (see Unit #1 Appendix), →Q remains on the menu until the menu is changed. Note that the accuracy of the conversion will depend on the display mode. Since the mode should be set to Std, the fractional conversion is accurate to 11 significant digits. Also round-off errors may affect the answer. See the example below.

Mixed Operations with Fractions

Example: $\left(\dfrac{2}{3}\right)^2 + \dfrac{1}{5}$

> **Keystrokes (Algebraic):** ['] [←] <( )> [2] [÷] [3] [►] [yˣ] [2] [+] [1] [÷] [5] [ENTER] [EVAL] (→Q)

Note that you do not get the right answer! This is because of round-off error affecting the accuracy of the fractional conversion. The decimal answer (before conversion) is displayed as 0.644444444445. If you enter 0.644444444444 then push (→Q), you will get the correct answer.

UNIT #3
SHORT CUT KEYS: HOW TO BE MORE STROKE EFFICIENT

Below is a description of some keys that you may find helpful as you explore the TI-82.

QUIT

You may quit any screen and return to the home screen by pressing [2nd] <QUIT>. This is helpful when you seem to be stuck on a screen and pressing [CLEAR] does not return you to your home screen.

INS

This key is helpful when you enter data incorrectly - particularly in expressions that are lengthy.

Enter the following expression on the calculator:

   3.24 + 6   DO NOT PRESS [ENTER]

Assume that you had meant to key in the number 3.214 rather than 3.24. You may insert the 1 in the appropriate place by first placing the cursor over the digit 4 by using the arrow keys. To insert the 1, press [2nd] <INS> [1]. The desired expression should now be displayed. Pressing [ENTER] gives us the sum of 9.214.

When using the insert key you should first place the cursor in the position that you want the inserted digit/symbol to appear, press [2nd] <INS> and then the desired digit/symbol. The calculator will insert as many characters as you like as long as you do not press a cursor key.

DEL

This key is helpful when you have entered a character that you do not want.

Enter the following expression on the calculator:

   6.542 X 9   DO NOT PRESS [ENTER]

Assume that you misread the number and simply meant to enter the number 6.5. To remove the 4 and the 2, place the cursor over the 4 by using the arrow keys.

Press [DEL] and the calculator deletes the digit 4. Pressing [DEL] again, deletes the digit 2. You now have the desired expression.

π

When evaluating formulas requiring the use of π, press [2nd] <π>. The calculator will evaluate the expression using a nine decimal approximation for π.

ENTRY

Pressing [2nd] <ENTRY> accesses the ability of the calculator to recall the expression previously entered.

Enter the following expression on the calculator, pressing [ENTER] to simplify. (This is the formula for the area of the trapezoid pictured below. Four is the height, 6 and 3 are the lengths of the parallel sides - called bases. Enter the arithmetic expression only.)

$$A = \left(\frac{1}{2}\right)(4)(6 + 3)$$

Record your answer here: _____

Assume you next wanted to compute the area of the trapezoid pictured whose bases are the same as our first trapezoid, but whose height differs.

Rather than re-entering the entire expression, we could press [2nd] <ENTRY>, retrieving our first expression, place the cursor over the 4, and enter our new height of 8. Try it!

This is particularly helpful when you mis-key an entry or just want to change some of the data.

[2nd] [◄]

These keystrokes move the cursor to the beginning of your entry.

28

[2nd] [▶]

These keystrokes move the cursor to the end of your entry.

ANS

Consider the problem below:

You have just gotten a new job and contracted for $525 weekly. How much money will you make in a year? (Consider a year to be comprised of 52 weeks).

Record your screen:

You will, however, be paid monthly. To figure your monthly salary you must divide the yearly salary just computed ($27,300) by 12. Press [÷] [1] [2] [ENTER], and record your screen below.

Notice when an operation sign is entered first, the calculator performs that operation with the answer from the previous entry.

The calculator will also recall the answer from the previous computation. You may access this function by pressing [2nd] <ANS> (ANS is located above the gray key used to enter a negative sign). The problem below demonstrates the use of this key.

Example: A farmer plants an apple orchard consisting of 38 rows of trees with 18 trees per row. How many trees are contained in the orchard? If his orchard produces 1197 bushels of apples, what is the average yield per tree?

Solution:

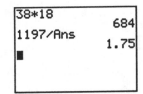

Therefore, the average yield is 1.75 bushels of apples per tree.

The EXIT key is used to remove a menu from the screen. For example, to remove the **CUSTOM** menu from the bottom of the screen, you simply press **[EXIT]**.

UNIT #3 APPENDIX
Casio fx7700GE

EXIT

This key is used to remove a menu from the screen. It is analogous to the **QUIT** key on the TI-82. For example, to remove a menu from the bottom of the screen, you simply press **[EXIT]**. Also, if you are unable to perform a calculation on a certain screen, you may try pressing **[EXIT]** to get to a computation screen.

REPLAY

There is no **<ENTRY>** key on the Casio. Instead, you can **REPLAY** your most recent formula by pushing the left arrow [◄] or right arrow [►] buttons. The right arrow brings you to the beginning of the line and the left arrow takes you to the end of the line. (Notice the word **REPLAY** engraved in the plastic between these two arrows.)

ANS

The **ANS** feature is accessed by pressing **[Shift] <ANS>** located above the [(-)] key.

31

UNIT #3 APPENDIX
HP 48G

There are a number of helpful keys for the HP 48G. Consult Chapter 2 of the User's Guide for entering and editing objects, Chapter 3 for manipulating the stack, and Chapter 7 for editing in the EquationWriter.

ON/CANCEL

Below the ON button is the word CANCEL written in white type. While the calculator is on, this is the function of the ON button. If you are entering data on the command line, pushing the CANCEL button will erase the data. If you are in an application such as the EquationWriter, CANCEL exits the application without executing.

###

The constant π is obtained from left-shifting the SPC button. π is displayed symbolically. To obtain the decimal approximation, press [←] <→NUM>.

UNDO

The HP 48G does not have an ENTRY button, but you can use UNDO to change an expression you have just evaluated. Use the algebraic-entry mode or the EquationWriter to enter an expression. Press [EVAL] to evaluate the expression. If you want to edit and reevaluate the expression, press [←] <UNDO> to undo the evaluation of the expression. Then press [←] <EDIT> to edit the expression. Then press [ENTER] [EVAL] to evaluate the new expression.

How do you simplify the expression $4 + 3 \cdot 2$? Do you work from left to right? Do you simplify the multiplication before the addition? The expression, seemingly, has more than one answer. This unit will help you construct an order of operations for simplifying arithmetic expressions.

Consider the above expression again. If you work from left to right, your answer is 14, as shown below:

$$4 + 3 \cdot 2 =$$
$$7 \cdot 2 =$$
$$14$$

If you perform the operation of multiplication first, your answer is 10:

$$4 + 3 \cdot 2 =$$
$$4 + 6 =$$
$$10$$

This unit will help you discover the agreement that exists on the simplification of arithmetic expressions that contain more than one operation.

1.   Simplify the following expressions using the TI-82. Record your answers on the blanks provided:

   4 - 3(2) _____   6(-2) - 8 _____   16 + 2(4) _____

   Place numbers above the operation sign or symbol to indicate which operation was performed first and which was performed second. Remember, the parentheses are indicators of the operation multiplication.

   When an expression contains the arithmetic operations of addition, subtraction, and multiplication, which operation does the calculator perform first? _____

2.   Simplify the following expressions by entering them on the calculator. Record your answers on the blanks provided.

   16 + 2 ÷ 2 _____   18 ÷ 3 + 5 _____   8 - 3 ÷ 5 _____

   Place numbers above the operation sign or symbol to indicate which operation was performed first and which was performed second.
   When an expression contains the arithmetic operations of addition, subtraction, and division, which operation is performed first? _____

3. Simplify each of the following expressions, recording the answer displayed by the TI-82 on the blank provided.
Place numbers above the sign of the operation to indicate which was performed first, second, third, etc.

a. $64 \div 8 - 3 \cdot 2$ _____

b. $12 \cdot 2 - 15 \div 3$ _____

c. $24 \cdot 2 + 3 \div 3$ _____

d. $16 \div 8 + 3 - 4 \cdot 2$ _____

e. $4 \cdot 2 - 3 + 16 \div 2$ _____

When all 4 arithmetic operations are contained in an expression, in what order does the calculator perform the operations?

4. The TI-82 performs the operations of multiplication and division before the operations of addition and subtraction. Moreover, it does the multiplication and division in the order in which they appear in the expressions (from left to right). It then performs the additions and subtractions in the order in which they appear from left to right.

To ensure you understand this order of operations, simplify the expression below in the space provided two ways. On the left, manually record each step; on the right, enter the given expression on the calculator and record the display.

"By hand"
$42 - 5 \cdot 3 + 4 \div 2$

"By Calculator"
$42 - 5 \cdot 3 + 4 \div 2$

If you did NOT get the same answer, go back and check both your manual computation and your TI-82 display (to ensure correct keystrokes).

5. The order of operations established in #4 can be overridden by using grouping symbols. Use the TI-82 to simplify each expression below, recording your answer on the blank provided.

   (4 + 2)3 _____   8(3 + 6) _____   24 ÷ (4 + 2) _____

   6(8-9) _____   60 ÷ (15-10) _____

   Put the number 1 over the operation that is performed initially. What does this tell us about the use of parentheses in the order of operations?

   Since parentheses are a <u>grouping symbol</u> where would grouping symbols be placed in our order of operations? Should it be placed above (before) multiplication and division or below (after) addition and subtraction?

   NOTE: Go back and look at your summary from Unit #1. Notice that the absolute value symbol and radical sign also serve as grouping symbols.

6. Simplify the expressions below using the TI-82. Record the displayed answer on the blank provided.

   $6 + 2^3$ _____ $24 ÷ 2^3$ _____

   $14 - 3^2$ _____ $6 \cdot 5^2$ _____

   Regardless of the arithmetic operation, does the calculator do the *power* first or the *arithmetic operation* first?_____

7. Simplify the expressions below using the TI-82. Record the displayed answer on the blank provided.

   $6 + (1 + 1)^3$ _____ $24 ÷ (3 - 1)^3$ _____

   $14 - (2 + 1)^2$ _____ $6(3 + 2)^2$ _____

   Which operation, or grouping symbol, did the calculator do first? _____
   Where would exponents be placed in our orders of operations list?

8. Simplify $\dfrac{8 + 16}{2}$ using the TI-82. Record your calculator screen.   ANS.12

   Does your screen indicate 8 + 16/2 or (8 + 16)/2?
   Explain, in your own words, why 8 + 16/2 **IS NOT** the correct format.

9. Simplify each of the expressions below using the TI-82. Record your
   calculator screen when you get the correct answer as indicated on the right.

   $\dfrac{6(4 + 5)}{3 + 2 \cdot 5}$ ANS. $\dfrac{54}{13}$

   $\dfrac{6(4 + 5)}{4 \div 2 + 1}$ ANS. 18

10. Look back at the display screens for the above problems. Notice how you had
    to enter the numerators and the denominators. What conclusion can you
    draw?

11. Enter the problem below on the calculator and record your screen display.

    5{8 + [3(4 - 2)]}

    If you enter brackets,[ ], and braces, { }, you will get an error message.
    Moreover, the calculator tells you that you made a SYNTAX error. This
    means that you entered a symbol inappropriately. On the TI-82, brackets and
    braces *are not* used as grouping symbols. ONLY PARENTHESES SERVE
    THAT FUNCTION.

    Re-enter the above expression, using parentheses only as grouping symbols.
    Before pressing [ENTER], it is a good idea to count the number of parentheses
    you have entered. There should be an even number. Insert any that you
    inadvertently omitted. Your answer should be 70.

12. Application: You invest $2,000 in an account that pays 4% interest per year. You leave the money in the account for 5 years. What amount of money is in the account at the end of the five year period?

Amount = Principal + Principal(Rate)(Time)

A = $2000 + 2000(.04)(5)

A = _____

Write your answer as a sentence:

13. Troubleshooting: Each of the problems below has been entered **incorrectly** on the calculator. Make the necessary corrections so that the calculator display accurately represents the problem given. Be sure to verify with your calculator!

a. $\dfrac{15 + -35}{42 - (-3)}$

b. $4 - 3\{6^2 - (2 - 9)\} + 5$

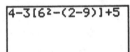

14. Summarizing Results: On the back of this page, summarize the order of operations you have learned in this unit. Write your summary in paragraph form. You should include the following points:
   a. a list of the order in which the calculator performs arithmetic operations (i.e. the hierarchy),
   b. the use of parentheses in fractions, and
   c. the manner in which grouping symbols are entered on the calculator.

Solutions: The correct keystrokes for #13 are:
a. [(] [1] [5] [+] [(-)] [3] [5] [)] [÷] [(] [4] [2] [-] [(-)] [3] [)]
b. [4] [-] [3] [(] [6] [x²] [-] [(] [2] [-] [9] [)] [)] [+] [5]

37

UNIT #4 APPENDIX
HP 48G

Use either the algebraic-entry mode or the EquationWriter to work the problems in this unit.

Variables are letters that are used to represent numbers. You have already encountered many variables in mathematics through the use of formulas. Variables allow us to represent relationships in a generalized format. For example, the formula for the area of a rectangle (A = lw) allows us to find the area for *any* rectangle, as long as we know its length and width. All that we need do is substitute the numerical values for the length and width of our rectangle into the formula, simplify the resulting arithmetic expression, and we have the area of our rectangle. Evaluating a formula means to find the value of one of the variables in the formula. Remember, the order of operations is in effect.

Evaluate each formula below by (a) substituting the given values into the formula (record the resulting arithmetic expression), and (b) entering the arithmetic expression you wrote in part (a) into the TI-82 appropriately.

Space is provided for you to record the arithmetic expression you enter on the calculator as well as to record your calculator screen *EXACTLY* as it appears. Answers are provided to ensure correct calculator entry.

Example: Evaluate $d = \sqrt{(x_2 - x_1)^2 + (y_2 - y_1)^2}$ when $x_1 = 8$, $x_2 = -3$, $y_1 = -1$, and $y_2 = 2$. Round your answer to the nearest hundredth.

Solution: a. $d = \sqrt{(-3-8)^2 + (2 - {}^-1)^2}$ b. screen display

```
√((-3-8)²+(2--1)
²)
          11.40175425
```

1. The formula for finding the slope of a line when given the coordinates of two

   points is $m = \dfrac{y_2 - y_1}{x_2 - x_1}$. Find m (the slope of the line) when $x_1 = 4$, $x_2 = 2$,

   $y_1 = -1$, and $y_2 = 5$. ANS. -3

   a. *arithmetic expression:*   b. *calculator display:*

41

Evaluate the slope formula when $x_1 = 6$, $x_2 = 9$, $y_1 = 4$, $y_2 = 6$.   ANS. $\frac{2}{3}$

a. *arithmetic expression:*   b. *calculator display:*

2.   The Pythagorean Theorem can be used to find the length of the hypotenuse of a right triangle. The formula can be expressed as $c = \sqrt{a^2 + b^2}$. Evaluate the formula when $a = 12$ and $b = 5$.   ANS. 13

a. *arithmetic expression:*   b. *calculator display:*

Evaluate the formula when $a = 4$ and $b = 9$. Round your answer to the nearest hundredth.   ANS. 9.85

a. *arithmetic expression:*   b. *calculator display:*

3.   The distance formula ($d = \sqrt{(x_2 - x_1)^2 + (y_2 - y_1)^2}$) can be used to find the distance between two points, (x_1, y_1) and (x_2, y_2), when their coordinates are known. Find d when $x_1 = -2$, $x_2 = 4$, $y_1 = 5$, and $y_2 = -3$. Round your answer to the nearest hundredth.   ANS. 10

a. *arithmetic expression:*   b. *calculator display:*

42

4. The quadratic formula can be used to find the value(s) of x in a quadratic

equation. One of the forms of the quadratic formula is $x = \dfrac{-b+\sqrt{b^2-4ac}}{2a}$.

Find x when a = 2, b = 6, and c = 3. Round your answer to the nearest hundredth. ANS. -.63

a. *arithmetic expression:*   b. *calculator display:*

Evaluate the formula when a = 1, b = -8, and c = 4. Again, round your answer to the nearest hundredth. ANS. 7.46

a. *arithmetic expression:*   b. *calculator display:*

5. The area of a circle is given by the formula $A = \pi r^2$. Recall that π is an irrational number (a constant) and "r" is the radius of the circle. Accessing π on the TI-82 gives a 9 decimal approximation of π. Use the above formula to find the area of a circle whose radius is 4". Round your answer to the nearest hundredth.
ANS. 50.27 square inches

a. *arithmetic expression:*   b. *calculator display:*

6. The volume of a sphere is given by the formula $V = \dfrac{4}{3}\pi r^3$. Find the volume of

a sphere whose radius is 6". Round your answer to the nearest hundredth.
ANS. 904.78 cubic inches

a. *arithmetic expression:*   b. *calculator display:*

7. The formula for the amount of money in a savings account at the end of a specified time is $A = p + prt$, where A is the amount of money, p is the principal (the initial amount invested), r is the rate (expressed as a decimal), and t is the time. If $2000 is invested in an account for 3 years earning 4% interest, what is the amount in the account at the end of that time?
ANS. $2240

    a. *arithmetic expression:*   b. *calculator display:*

8. $S = \dfrac{n(a + L)}{2}$ is the formula for the sum of n terms in an arithmetic sequence of numbers. Use this formula to find the sum (S) of the first 100 natural numbers. Let $a = 1$ (the first number in the sequence), $L = 100$ (the last number of the sequence), and $n = 100$ (the number of numbers in the sequence).
ANS. 5050

    a. *arithmetic expression:*   b. *calculator display:*

Use the above formula to find the sum of the even natural numbers from 2 to 100. First, identify the value for each of the variables in the formula.

$a =$ _____ (the first number in the sequence)

$L =$ _____ (the last number in the sequence)

$n =$ _____ (the number of numbers in the sequence)

ANS. 2550

    a. *arithmetic expression:*   b. *calculator display:*

9. The formula used to convert degrees Fahrenheit to degrees Centigrade is

    $C=\frac{5}{9}(F-32)$. If the outside temperature is 85° Fahrenheit, what is the
    equivalent Centigrade temperature? (Round to the nearest degree).
    ANS. 29°

    a.*arithmetic expression:*   b.*calculator display:*

10. An algebra student made the following grades on the 4 tests given before the
    final exam: 85, 92, 73, 88. Write a formula that will enable any student in
    this class to compute the average of *their* four tests. Use the notation of T_1
    for the variable representing the grade on the first test, T_2 for the variable
    representing the grade on the second test, T_3 to represent the grade on the
    third test, and T_4 for the fourth test grade. Use $\bar{x}$ as the variable
    representation of the average.

    Formula:

    Use your formula to compute the above student's average. ANS. 84.5

    a.*arithmetic expression:*   b.*calculator display:*

11. Application: A newly licensed realtor is offered a job at 2 real estate
    companies. Hearth and Home offers a salary of $500 per month and pays a
    commission of 3% on each sale (Income = 500+.03s). The Homes 4 U
    Company offers a flat percentage of 4% on sales (Income = .04x). If the
    newly licensed realtor sells a house valued at $124,000, which company
    would pay the higher salary?

12.	Summarizing Results: Write a summary of what you have learned in this unit. Look back at the screens you recorded and compare them to your hand-written arithmetic expressions. Your summary should include a discussion of the use of parentheses in the entering of expressions containing radicals and fractions.

UNIT #5 APPENDIX
HP 48G

Use either the algebraic-entry mode or the EquationWriter to work the problems in this unit.

UNIT #6
INTRODUCTION TO POSITIVE AND NEGATIVE EXPONENTS

In Unit #1 you learned how to enter positive exponents on the graphing calculator. Positive exponents are a quick way to write repeated multiplications of the same factor. Negative exponents cannot be used to represent repeated multiplication because $x^n = x \cdot x \cdot x \cdot x \ldots \cdot x$ (for n factors of x) is defined in terms of n being a counting number.

REMEMBER: The " $\wedge$ " key is used to raise a number to a power. If your number is raised to the second power, as in 5^2, you have the option of keystroking [5] [x^2] or [5] [$\wedge$] [2]. However, for powers other than 2 you must use the " $\wedge$ " key.

1. Enter each of these problems on your TI-82. Record your screen display and enter the result on the blank.

    a. 3^1 a._____

    b. 3^2 b._____

    c. 3^3 c._____

    d. 3^4 d._____

    e. 3^5 e._____

2. As the exponent is increased on a positive base number, what can you conclude will happen to your result?

3. As the exponent is increased on a negative base number, what would you predict will happen?

4. Try these (following the directions in #1) to see if your prediction was accurate. Be sure to enter the following expressions as written - **include** parentheses.

    a. $(-3)^1$ a._____

    b. $(-3)^2$ b._____

49

c. $(-3)^3$ c._____

d. $(-3)^4$ d._____

e. $(-3)^5$ e._____

5. Clearly describe what happens as a negative base is raised to a positive increasing power.

6. What will happen if we raise a positive base to a negative power? Follow the same directions as in #1 and observe what happens. Answers will be displayed as decimals.

     a. 3^{-1} a._____

     b. 3^{-2} b._____

     c. 3^{-3} c._____

     d. 3^{-4} d._____

     e. 3^{-5} e._____

7. Rework each of the problems in #6 (with the calculator!) and convert to a fraction before you press [ENTER]. The first one has the keystrokes displayed for you.

     a. 3^{-1} [3] [∧] [(-)] [1] [MATH] [ENTER] (remember,   a._____
                        you are selecting option 1) [ENTER]

     b. 3^{-2} b._____

     c. 3^{-3} c._____

     d. 3^{-4} d._____

     e. 3^{-5} e._____

8. Follow the same directions as in #1, but be sure to use the ▶FRAC option to convert all answers to fractions (as was done in #7).

     a. $\dfrac{1}{3^1}$ a._____

     b. $\dfrac{1}{3^2}$ b._____

c. $\dfrac{1}{3^3}$ c._____

d. $\dfrac{1}{3^4}$ d._____

e. $\dfrac{1}{3^5}$ e._____

9. In your own words, describe what effect the negative exponent has on the base number. (Compare #7 and #8 to help you out.)

10. Follow the same directions as in #1 and simplify the following:

a. $\left(\dfrac{1}{3}\right)^{-1}$ a._____

b. $\left(\dfrac{1}{3}\right)^{-2}$ b._____

c. $\left(\dfrac{1}{3}\right)^{-3}$ c._____

d. $\left(\dfrac{1}{3}\right)^{-4}$ d._____

e. $\left(\dfrac{1}{3}\right)^{-5}$ e._____

11. Was your conjecture in #9 about the effect of the negative exponent still valid after you completed #10?

Compare these two answers: 3^{-1} = _____ $\left(\dfrac{1}{3}\right)^{-1}$ = _____

What effect does the negative exponent have?

12. Each of the following problems should be worked in two ways:
1) with the calculator, copying your screen display to justify your work (before computing the result, use the ▸FRAC option from the MATH menu so that all answers that are fractions will be displayed as such), and
2) by hand, using the Laws of Exponents - **SHOW** each step of your work. Be sure this "hand computation" agrees with your "calculator computation".

a. $2^{-3} \cdot 2^3$

"By Hand":

Calculator display:

ANS. 1

b. $\dfrac{2^{-3} \cdot 2^3}{2^3}$

"By Hand":

Calculator display:

ANS. $\dfrac{1}{8}$

c. $2^{-3} + 2^3$

"By Hand":

Calculator display:

ANS. $\dfrac{65}{8}$

d. $\dfrac{2^{-1}}{5^{1}}$

"By Hand":

Calculator display:

ANS. $\dfrac{1}{10}$

e. $\dfrac{5^{1}}{2^{-1}}$

"By Hand":

Calculator display:

ANS. 10

f. $\dfrac{2^{-1}}{5^{-1}}$

"By Hand":

Calculator display:

ANS. $\dfrac{5}{2}$

g. $\dfrac{5^{-1}}{2^{-1}}$

"By Hand":

Calculator display:

ANS. $\dfrac{2}{5}$

13. Troubleshooting: Each of the problems below has been entered **incorrectly** on the calculator. Make the necessary corrections so that the calculator display accurately represents the problem given. Be sure to verify with your calculator!

a. $(-5)^4$

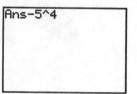
```
Ans-5^4
```

b. -6^6

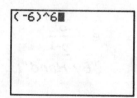
```
(-6)^6■
```

c. $\left(\dfrac{2}{3}\right)^4$

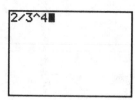
```
2/3^4■
```

14. Summarizing Results:
   a. Discuss the effect that a negative exponent has on the base. You should address bases that are both integers and fractions.
   b. Why was the zero exponent not addressed in this unit? (HINT: Refer back to title.)
   c. What effect does the zero exponent have on the base?

15. Highlight the REMEMBER note in the introductory paragraph to this unit for your QUICK REFERENCE list of operations.

<u>Solutions:</u> The correct keystrokes to #13 are:
a. [(] [(-)] [5] [)] [∧] [4]　　b. [(-)] [6] [∧] [6]　　　　c. [(] [2] [÷] [3] [)] [∧] [4]

Unless you are working entirely in fractions, the Casio does not have the capability to convert a *decimal* into a *fraction*. You may want to make a table of common fractions together with their decimal expansions so that when you see the decimals on your screen, you can give the exact fractional answer.

Example: 0.333333333333 = 1/3

0.125 = 1/8

In #4 it is not necessary to use parentheses if you are using the stack. If you are using algebraic-entry mode or the EquationWriter be sure to include parentheses.

In #7 you can only convert to fractions using (→Q) after first computing the decimal answer. Beware of any round-off errors (see Unit#2 Appendix).

UNIT #7
SCIENTIFIC NOTATION

Multiplying numbers by 10 or powers of 10 produces an interesting pattern.

1. Use the TI-82 to simplify each of the expressions below. The "X" used is a multiplication sign - *not* a variable.

$$2.56 \times 10^2 = \underline{\hspace{3cm}}$$

$$3.5 \times 10^3 = \underline{\hspace{3cm}}$$

$$6.2 \times 10^4 = \underline{\hspace{3cm}}$$

$$8 \times 10^5 = \underline{\hspace{3cm}}$$

Compare the placement of the decimal in the original number to its placement in the value of the expression entered on the given lines. Notice how the decimal always moves to the right. Now compare the number of places the decimal moves to the power of 10 specified in the original expression. Do you notice a pattern? Describe that pattern in the space below.

2. Use the TI-82 to again simplify each of the expressions below. Again, the "X" is used is a multiplication sign and *not* as a variable.

$$3.658 \times 10^{-1} = \underline{\hspace{3cm}}$$

$$2 \times 10^{-2} = \underline{\hspace{3cm}}$$

$$7.2 \times 10^{-3} = \underline{\hspace{3cm}}$$

Compare the placement of the decimal in the original number to its placement in the value of the expression. Notice how the decimal always moved to the left. Now compare the number of places the decimal moved to the power of 10 specified in the original expression. Do you notice a pattern? Describe that pattern in the space below.

3. Using your results from #1 and #2 above, simplify each of the following expressions *without* using your calculator. You should be able to do them using the method you developed above.

$$4.65 \times 10^4 = \underline{\hspace{3cm}}$$

$$6.3 \times 10^3 = \underline{\hspace{3cm}}$$

$$7.658 \times 10^5 = \underline{\hspace{3cm}}$$

$$6.2 \times 10^{-2} = \underline{\hspace{3cm}}$$

$$9.5 \times 10^{-5} = \underline{\hspace{3cm}}$$

Check your work by simplifying the expressions using the TI-82.

Each of the numbers above is written in scientific notation. This is a notation used by scientists to write very large or very small numbers. When writing a number in scientific notation, notice that the first number is ALWAYS greater than or equal to 1 and less than 10. The power of ten indicates the number of places the decimal is moved - movement is to the left when the exponent is negative, to the right when the exponent is positive.

4. The number 46.678×10^2 **IS** written as the product of a number and a power of 10, but **IS NOT** written in scientific notation. Why not?

    Rewrite the number correctly in scientific notation: $\underline{\hspace{3cm}}$

5. Write each of the numbers below in scientific notation:

    456,000 $\underline{\hspace{4cm}}$

    0.0000000000789 $\underline{\hspace{4cm}}$

    6,789,000,000,000 $\underline{\hspace{4cm}}$

    0.03 $\underline{\hspace{4cm}}$

6. The TI-82 has a key labeled **MODE** which will enable us to check our answers. After turning the calculator on, press **[MODE]**. All of the options on the left should be highlighted. We want to change the setting on the first line from **NORMAL** to **SCI** (scientific). To do this, use the right cursor arrow to place the blinking cursor over **SCI**. To choose this mode, press **[ENTER]**. Your screen should now look like this.

    ```
    Normal SCI Eng
    Float 0123456789
    Radian Degree
    Func Par Pol Seq
    Connected Dot
    Sequential Simul
              Split
    ```

Press [CLEAR] to return to the home screen.

7. Now enter the numbers below (one at a time) into the calculator. You should press [ENTER] after each entry. Copy the displayed answer.

456,000 _____

0.0000000000789 _____

6,789,000,000,000 _____

0.03 _____

8. In #5, you should have written 456,000 as 4.56×10^5 in scientific notation. However, in #7, the calculator displayed that number as **4.56 E 5**. The notation the calculator displays differs from the notation we use. Explain/reconcile the difference.

9. Complete the chart below by entering the number on the left in the TI-82, recording the display, and then writing the number in scientific notation. The first one has been done for you to provide a model.

Standard Notation	Calculator Display	Scientific Notation
8,000	8 E 3	8×10^3
0.00358		
2,000,000		
0.0124		
67,300		

10. Complete each computation below by entering each expression on the calculator and pressing [ENTER]. Record your answer in *scientific notation* (not, however, in the calculator's display of scientific notation).

a. (0.006)(3.5987) _____

b. $\dfrac{(6,000,000)(40,000)}{3,000}$ _____

c. $\dfrac{(0.0000008)(5,000,000)}{(0.0004)(0.00005)}$ _____

61

11. Convert your answers above to standard notation by either using the rules you stated in #2 OR by resetting the calculator back in **NORMAL** mode. To do this, press **[MODE] [ENTER]**. Return to your home screen by pressing **[CLEAR]**. Check your answers in #10 above as you did in #1.

NOTE: Even when NOT in SCI mode, the calculator will express very large or very small numbers in scientific notation.

For example: With the calculator in NORMAL MODE, simplify the following expression using the TI-82. Record the calculator display in the space below:
        (0.00006)(0.0000008)

Write your answer in scientific notation (not calculator notation!) _____

Write your answer in standard notation: _____

12. Summarizing Results: Summarize what you learned in this unit. Your summary should address:
    a. converting a number from standard notation to scientific notation (both with the calculator and without it), and
    b. converting a number from scientific notation to standard notation (both with the calculator and without it).

13. Highlight #6 and #8 in this unit for your QUICK REFERENCE list of operations.

UNIT #7 APPENDIX
Casio fx7700GE

To work entirely in scientific notation on the Casio, select [SHIFT] <DISP> [F2] (Sci) for Sci then choose an appropriate number of significant digits and press [EXE]. (For most applications, 4 significant digits is plenty.)

To return to standard notation, select [SHIFT] <DISP> [F3](Nrm).

UNIT #7 APPENDIX
HP 48G

To display numbers in scientific notation, go to the CALCULATOR MODES directory by pressing [→] <MODES>. Move the cursor such that Std is highlighted. Press (CHOOS) then move the cursor over Scientific and press [ENTER]. Then press [►] so that the cursor is over 0 and then enter the number of decimal places you with to have displayed. Four should be sufficient.

To return to standard display, return to the CALCULATOR MODES directory and choose Std.

UNIT #8
USING THE STOre KEY TO EVALUATE EXPRESSIONS AND CHECK SOLUTIONS

The TI-82 can be used to evaluate expressions. To "evaluate an expression" means to find the **VALUE** of the expression for assigned values of the variable. Evaluating with a calculator can be accomplished in two ways. First, you can simply substitute the values in by hand and then enter the resulting expression in the calculator. This approach was examined in Unit 6. The second approach is to store the values under different variable names by using the **STO▸** key.

1. To store a value, simply enter the value, press [STO▸] and then the desired variable. For example, to store X = 5, press [5] [STO▸] [X,T,θ] [ENTER]. The value 5 is now stored under the variable X. This value will remain stored in X until another value replaces it. If you want to check to see what value is actually stored under a specific variable, enter the variable at the prompt (blinking cursor) and then press [ENTER]. The value is then displayed on the screen.
    NOTE: Because "X" is used as a variable so often in algebra, it has its own key on the TI-82. We are in function MODE; therefore when we press [X,T,θ] the screen displays the variable "X".

2. To store a value under a variable other than "X", you must use the [ALPHA] key to access the 26 letters of the alphabet. Pressing the [ALPHA] key, followed by another key, allows you to access the upper case letters and symbols written in white above the key pads.

3. **Evaluate $3X^2 + 6X + 2$ when X = -11.**

    Solution: Store X = -11, by pressing [(-)] [1] [1] [STO▸] [X,T,θ]. The colon ":" key (located above the gray decimal key) is used to separate commands that are to be entered on the same line, so we will now press [2nd] <:> before telling the calculator what expression to evaluate. Press [3] [X,T,θ] [x²] [+] [6] [X,T,θ] [+] [2] to enter the expression that is to be evaluated. At this point, we have told the calculator to store 11 in X and evaluate $3X^2 + 6X + 2$ at this value of X.

    ```
    -11→X:3X²+6X+2
                  299
    ■
    ```

    The calculator will not perform the instructions until you press [ENTER]. The polynomial evaluates to 299. Your screen should look like the one above.

4. **Evaluate** $a^2 - 3a + 5$ when $a = -7$.

Solution: See screen at right.

(HINT: Refer back to #2 and #4 if you need keystroke assistance.

```
-7→A:A²-3A+5
                    75
■
```

NOTE: Be sure to use the gray [(-)] key for the negative sign on a number and use the blue [-] key to indicate subtraction.

5. **Evaluate** $X^2 - 2XY + Y^2$ when $X = -2$ and $Y = 3$.

Solution: Look at the screen displayed at the right. What have you told the calculator to do?

```
-2→X:3→Y:X²-2XY+
Y²
                    25
```

-2→X means _____

3→Y means _____

$X^2 - 2XY + Y^2$ means

_____

25 means

_____

The purpose of the colon (:) is to_____

_____

EXERCISE SET

Do each of the problems A - E below in two ways:
(i) paper and pencil, showing each step of your work and
(ii) with the calculator, using the **STO►** feature. Record the screen display as a means of showing your work. You must copy the screen display **exactly** as it appears, line by line.

A.   Evaluate $2X^2 + 3X + 1$
          when $X = 2$ ANS. 15
"By Hand":   *Calculator display:*

when X = -3   ANS. 10
"By Hand":   Calculator display:

B.   Evaluate $(X - 2)^2 + 3X$
          when X = ½   ANS. $\dfrac{15}{4}$
"By Hand":   Calculator display:

when X = -2   ANS. 10
"By Hand":   Calculator display:

C.   Evaluate $4(X - 3) + 5(Y - 3) - 4$
          when X = -1 and Y = -3   ANS. -50
"By Hand":   Calculator display:

67

D.   Evaluate $-X^2 + 3XY^2 - 6Y^3$
        when X = 3 and Y = 4   ANS. -249

"By Hand":   *Calculator display:*

E.   Evaluate $(X - 5)^2 + 2XY^2 - 6$
        when X = 2 and Y = -3   ANS. 39

"By Hand":   *Calculator display:*

6.   Troubleshooting: Each of the problems below has been entered **incorrectly** on the calculator. Make the necessary corrections so that the calculator display accurately represents the problem given. Be sure to verify with your calculator!

a. Evaluate 2X - 5 when X = -3.

```
X→-3:2X-5█
```

b. Evaluate $\dfrac{\sqrt{B^2 - 4AC}}{2A}$ when B = 4, A = 5, C = -1.

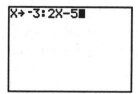

```
5→A:4→B:-1→C:√B²
-4AC
```

7. Is 6 a solution to the equation $4(2X - 1) - 8 = 42 - X$?

```
6→X:4(2X-1)-8
                36
42-X
                36
```

    Solution: If 6 is a solution to the equation then the left
    side, $4(2X - 1) - 8$, will have the same value as the right
    side, $42 - X$, when 6 is substituted for X. Store 6 in X and
    then find the value of each side of the equation.
    Your screen should look like the one at the above right.

8. Simplify the following expression algebraically:
    $3X(X - 2) + 5X^2$

9. Your graphing calculator can help you determine if your simplification is
    correct. Select a value for X and evaluate both the
    problem and your simplification. If both expressions have
    the same value then you know they are equivalent. You

```
5→X:3X(X-2)+5X²
                170
8X²-6X
                170
■
```

    must do this procedure for two different X values to
    ensure equivalent expressions. In order for everyone's
    problem to look the same, let X = 5. Store 5 in X and
    evaluate both the problem and the answer you determined. Your screen will
    look like the one at the above right.
    To check a second value, let X = 2 and evaluate both the original expression
    and the simplified expression. In both cases you should get a value of 20.

    WARNING: This procedure tells you that your simplification is equivalent to
    the original expression. It <u>does</u> <u>not</u> determine if your expression is completely
    simplified.

EXERCISE SET CONTINUED

Do each of the problems below by using the TI-82. Remember to use the
STO► feature. Record the screen display as a means of showing your work.
You must copy the screen display <u>**exactly**</u> as it appears, line by line.

F.   Is -9/2 a solution to $3X - 2 = 5X + 7$?

G.   Is -3 a solution to 3(X - 1) + 1 = 2X - 5?

NOTE: These techniques can be used to check all types of equations. They may also be used to check any problem in which you are simplifying algebraic expressions or factoring algebraic expressions.

H.   The expression $(3X^2 - 5X + 2) - (8X^2 - 3X - 1)$ simplifies to $-5X^2 - 2X + 3$. Check that these two expressions are equal by evaluating both the original expression and its simplified form when X = 4 and when X = 3.

I.   The expression $(4X - 5)(3X + 7)$ expands to $12X^2 + 13X - 35$. Check that these two expressions are equal by evaluating both the original expression and its simplified form when X = -2 and when X = 1.

J.   In attempting to perform the division on the expression $\dfrac{4x^2 - 8x - 16}{8x^2}$,

a student produced $\dfrac{1}{2} - \dfrac{1}{x} - \dfrac{4}{2x^2}$.

a. When he tried to use the calculator to check his work he selected 0 as his value for X and the calculator gave him an error message. What did he do wrong?

70

b. Next he tried selecting 5 for X and was relieved to discover that both the expressions evaluated to the same value. Does the calculator now guarantee him that his problem is completely finished? Explain.

NOTE: You can determine the value stored in the calculator for any given variable by pressing [ALPHA] <variable letter> [ENTER]. This retrieves the value stored in that variable. For example, press [ALPHA] <A> [ENTER] to see the value store in A. Your calculator should display "-7". (The TI-82 uses only uppercase letters.) Now repeat this procedure to determine the stored values for the following variables:

$Q = $ _____ $V = $ _____ $Z = $ _____

(Values for Q, V, and Z may vary from calculator to calculator. All three may even have the same value stored in them.)

K.   Application: The formula for the area of a rectangle is A = LW. Find the area of the following rectangles.

Rectangle 1: L = 3, W = 2   A = _____

Rectangle 2: L = 6, W = 2   A = _____

Rectangle 3: L = 12, W = 2   A = _____

What effect does doubling the length have on the area of the rectangle?

(If you entered three separate problems, return to Unit #3 for a reminder of how to shortcut your work.)

10. Summarizing Results: Write a summary of what you have learned in this unit. You should address the following:
    a. use of the [X,T,θ] key,
    b. storing a value for a variable,
    c. use of the colon on a command line,
    d. evaluating an expression with the STO► key,
    e. verifying that two expressions are equal, and
    f. checking roots of equations.

11. Highlight #1 and the NOTE after "J" in this unit for your QUICK REFERENCE list of operations.

Solutions: 6a. Your screen display should indicate that -3 is stored in X: -3→X
6b. After the values are stored in the given variables, the display for the expression should look like the following: $\sqrt{(B^2 - 4AC)} / (2A)$

There is no [X,T,θ] key on the TI-85. Instead, you will use the [x-VAR] key. In the TI-85, the variable is displayed as "x", instead of "X" as in the TI-82. Therefore when using this text, each "X" printed as a variable for the TI-82 should appear as "x" using the TI-85.

To store a value under a variable other than x you will not need the **ALPHA** key as the TI-82 does. Pressing **STO▸** automatically initiates the alpha key. (When you press **STO▸** you should notice that the blinking cursor has an A in it for **ALPHA.**

NOTE: Since the **STO▸** key automatically initiates the ALPHA key, you must press [ALPHA] to "turn off" the ALPHA key before you can access non-alphabetic keys.

In Exercise Set K, it should be noted that the TI-85 will not recognize LW as L times W unless you indicate the multiplication symbol, L*W. (The calculator will give you an ERROR 14 UNDEFINED if you enter LW instead of L*W.) The TI-85 has a much larger range of variables than does the TI-82. The TI-82 recognizes the 26 uppercase letters of the alphabet as storage locations. The TI-85 recognizes both uppercase and lowercase letters as separate variables. For example, w and W are as different as X and Y. In addition, the TI-85 recognizes combinations of letters as variables. Thus if you wanted to store a temperature value of 32° for a specific formula, you could enter the variable as TEMP or temp.

Example: Find the area of a rectangle whose length is 5 and whose width is 7.

Solution: The key strokes would be: [5] [STO▸] <L> [ALPHA] [2nd] <:> [7] [STO▸] [W] [ALPHA] [2nd] <:> [ALPHA] <L> [x] [ALPHA] <W> [ENTER]

On the Casio, the [→] key corresponds to the [STO] key on the TI-82.

1. To store a value, enter the desired value on level 1 of the stack. Then type the name of the variable. (Consult Chapter Five of the User's Guide for guidelines on proper variable names.) Then push **[STO]** to store. The variable can be displayed on the VAR menu. Push **[VAR]** to display the variable menu. For example, to store X = 5, press **[5] [ENTER] [α] <X> [STO]**. If you want to check to see what value is stored, activate the VAR menu by pushing **[VAR]** then select the menu key, **(X)**, then push **[ENTER]**. The value of X will be displayed on level 1. To change the value of a variable, enter the new value onto the stack then press **[←]** followed by the variable's menu key. For example, to change X = 5 to X = 4, press **[4] [ENTER] [←] (X)**.

3. **Evaluate:** $3X^2 + 6X + 2$ when X = -11

    Solution: Store X = -11. If the variable X already exists, press **[11] [+/-] [ENTER] [←] (X)**. Then enter expression onto the first level of the stack using algebraic-entry mode by pressing **['] [3] (X) [yˣ] [2] [+] [6] (X) [+] [2] [ENTER]**. Press **[EVAL]** to evaluate the expression.

UNIT #9
RATIONAL EXPONENTS AND RADICALS

Radicals

The inverse of raising to a power is extracting a root. For example, $4^3 = 64$ and $\sqrt[3]{64} = 4$.

1.   The only type of radical that has been addressed thus far in this text is $\sqrt{}$ or principal square root. We will now examine $\sqrt[3]{}$, $\sqrt[4]{}$, $\sqrt[6]{}$, etc., and the relationship of these radicals to rational exponents.

2.   We begin by looking at the MATH menu. Press [MATH] and the screen at the right is displayed. Pressing the down arrow key will allow you to scroll through the entire menu selection. This unit will use the ►FRAC option, as well as the $\sqrt[3]{}$ and $\sqrt[x]{}$ options.

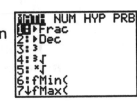

3.   If we wish to find the cube root of -27, $\sqrt[3]{-27}$, recall we are looking for a number that when raised to the third power yields -27. Press [MATH] [4] (to select the $\sqrt[3]{}$ option) [(-)] [2] [7] [ENTER]. The result displayed should be -3.

4.   To compute seven times the cube root of 5, $7\sqrt[3]{5}$, press [7] [MATH] [4:$\sqrt[3]{}$] [5]. The result displayed is an approximate answer rounded to nine decimal places. Answer: 11.96983163

5.   To compute roots other than square roots ($\sqrt{}$ is found on the face of the calculator) or cube roots, you must use the $\sqrt[x]{}$ selection on the MATH menu. Designate a value for x by entering the root value (or index) first then the $\sqrt[x]{}$ symbol.

6.   To compute the sixth root of 64, $\sqrt[6]{64}$, press [6] [MATH] [5:$\sqrt[x]{}$] [6] [4] [ENTER]. Compare your screen to the one at the right.

77

7. Because the display is $6\sqrt[x]{64}$, you must be careful when entering problems like $4\sqrt[6]{64}$. In #4 you did not have to "tell" the calculator to multiply 7 times $\sqrt[3]{5}$ because $7\sqrt[3]{5}$ indicates <u>implied</u> multiplication. However, in $4\sqrt[6]{64}$ you cannot use implied multiplication because of the calculator notation. If you

enter $4\ 6\sqrt[x]{64}$ the calculator will compute $\sqrt[46]{64}$. A multiplication symbol, *, must be entered after the 4 to clarify the problem for the calculator. Your screen display for $4\sqrt[6]{64}$ should look like the one at the right.

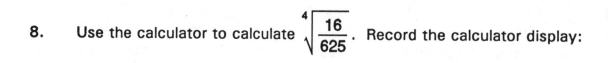

```
4*6 *√64
                8
```

8. Use the calculator to calculate $\sqrt[4]{\dfrac{16}{625}}$. Record the calculator display:

What is $\sqrt[4]{\dfrac{16}{625}}$ according to the calculator?_____ You should have 2/5 (or 0.4). If you got 2/625 (or 0.0032) then you failed to group the fraction with parentheses. Parentheses are necessary for the calculator to know that you are asking for the fourth root of the <u>quantity</u> 16/625.

9. Evaluate each radical expression. Record your screen display, being sure that the answer is displayed as an integer or fraction - not a decimal.

   a. $\sqrt[3]{-125}$ ANS. -5

   b. $\sqrt[4]{4096}$ ANS. 8

   c. $\sqrt[6]{5^6}$ ANS. 5

   d. $\sqrt{\dfrac{4}{25}}$ ANS. $\dfrac{2}{5}$

e. $\sqrt[5]{\dfrac{1}{32}}$

ANS. $\dfrac{1}{2}$

f. $\sqrt[4]{-625}$

ERROR Message

10. Did you get an error message on 9f? The DOMAIN error message is displayed because the TI-82 only computes values of real numbers. In the real number system, $\sqrt[4]{-625}$ does not exist. There is no real number that when raised to the fourth power will equal -625.

Rational Exponents

Positive exponents increase the size of a whole number and negative exponents decrease the size. You discovered in the unit on positive and negative exponents that negative exponents have the effect of taking the reciprocal of a number. What effect will rational (fractional) exponents have?

11. Rational exponents are entered into the calculator in the same manner as positive and negative integer exponents. You must, however, be careful of the rules for order of operations. A problem has been entered on the calculator and is displayed on the screen at the right. Two operations are indicated, exponent and division. Recalling the order of operation rules, fill in the operations in the order that they will be performed.

`25^1/2`

a._____ b._____

12. The previous screen demonstrates a typical entry error for the problem $25^{1/2}$. It is important to remember to place parentheses around the rational exponent. Your screen entry for $25^{1/2}$ should look like the one at the right.

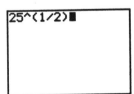

`25^(1/2)`

13. The following problems have been placed in groups of four so you can compare and contrast the effect of the exponents. **Be sure all rational exponents are enclosed in parentheses.** Record your screen display beneath each problem and enter the final result on the blank.

a. $25^2 =$ _____ $25^{1/2} =$ _____ $25^{-1/2} =$ _____ $25^{-2} =$ _____

b. $8^3 =$ _____ $8^{1/3} =$ _____ $8^{-1/3} =$ _____ $8^{-3} =$ _____

c. $\left(\dfrac{16}{81}\right)^2 =$ _____ $\left(\dfrac{16}{81}\right)^{\frac{1}{2}} =$ _____ $\left(\dfrac{16}{81}\right)^{-\frac{1}{2}} =$ _____ $\left(\dfrac{16}{81}\right)^{-2} =$ _____

14. If you scan through the chapter in your textbook that deals with rational exponents you will discover that the bulk of the chapter is dedicated to radicals, not rational exponents. This unit will conclude by comparing radical expressions to rational exponential expressions. You the student are left to determine the relationship between $\sqrt[n]{a^m}$ and $a^{m/n}$ by working the following problems. The problems are grouped so that you can more easily compare patterns. Copy your screen display beneath each problem and record the result on the blank.

a. $\sqrt[3]{4^6} =$ _____ $4^{\frac{6}{3}} =$ _____ $4^2 =$ _____

b. $\sqrt[3]{-8} =$ _____ $(-8)^{\frac{1}{3}} =$ _____ $\sqrt[3]{(-2)^3} =$ _____

80

c. $\sqrt{16} =$ _____ $16^{\frac{1}{2}} =$ _____ $\left(\frac{1}{16}\right)^{-\frac{1}{2}} =$ _____

d. $27^{\frac{2}{3}} =$ _____ $\left(\sqrt[3]{27}\right)^2 =$ _____ $\sqrt[3]{27^2} =$ _____

e. $\sqrt{-36} =$ _____ $(-36)^{\frac{1}{2}} =$ _____

15. In 14e. your calculator displayed a DOMAIN ERROR. Explain what is wrong with the problem.

16. Summarizing Results: Summarize what you have learned in this unit. You should address the following:
a. extracting roots other than square roots, and
b. the relationship between rational exponents and roots.

17. Highlight #2, #6, #7 and #8 of this unit for your QUICK REFERENCE list of operations.

Solutions: **13a.** 625, 5, 1/5, 1/625   **13b.** 512, 2, 1/2, 1/512   **13c.** 256/6561, 4/9, 9/4, 6561/256   **14a.** 16, 16, 16   **14b.** -2, -2, -2   **14c.** 4, 4, 4   **14d.** 9, 9, 9   **14e.** not a real number, not a real number

There is no $\sqrt[3]{}$ available on the TI-85. The $\sqrt[x]{}$ however is available under the (MISC) submenu of MATH ([2nd] <MATH> [F5] (MISC) [MORE] [F4] ($\sqrt[x]{}$)). You should add $\sqrt[x]{}$ to your CUSTOM menu (using <CATALOG>) as directed in Units #1 and #2 Appendix TI-85, or see the TI-85 GUIDEBOOK for instruction.

In reference to step 15: The TI-85 will not display a domain error. The TI-82 computes only with Real numbers; the TI-85 will perform complex number operations. When you enter $\sqrt{-36}$ on the TI-85, it will return (0,6). This is the calculator's notation for complex numbers. (0,6) = 0 + 6i

Radicals

On the Casio, the $< \sqrt[3]{} >$ key is located above the [( ] key and the $< \sqrt[x]{} >$ key is located above the [^] key.

Radicals

To take radicals use the $\sqrt[x]{}$ key which the right-shift of the $\sqrt{x}$ button. If you are using the stack, enter the number, enter the root value, then press [→] $<\sqrt[x]{}>$. If you are using algebraic-entry mode, press [→] $<\sqrt[x]{}>$, enter the root value, press [←] $<,>$, enter the number, then press [ENTER]. Press [EVAL] to evaluate.

CORRELATION CHART

UNIT * Instructors are encouraged to use Units marked by an * to introduce concepts.	PRE-REQ. UNIT(S)	CORRELATING CONCEPT
#10 Graphical solutions to Linear Equations	#1-4, #8	Solving linear equations
#11 Linear Applications (Interest, Profit, Breakeven points, Depreciation)	#10	Finding and interpreting solutions to linear equations through the use of graphs and tables.
#12 Graphical Solutions to Linear Inequalities	#10	Linear inequalities
#13 Graphical Solutions to Absolute Value Equations *	#10	Absolute value equations
#14 Graphical Solutions to Absolute Value Inequalities *	#10, #12	This unit can be used to allow the student to "discover" the algorithms typically used for solving.
#15 Solving Factorable Polynomial Equations *	#1-3, #7, #11	This unit can be used to introduce solving quadratic and other polynomials by factoring.
#16 Solving Non-factorable Quadratics	#10, #15	Solving non-factorable quadratic equations graphically
#17 Applications of Quadratic Equations *	#16	Interpreting graphs of quadratic equations in real life situations
#18 Solving Quadratic Inequalities *	#16	Solutions of quadratic inequalities are determined through graphs and tables.
#19 Solving Nonlinear Inequalities *	#18	Solutions of nonlinear inequalities are determined through graphs and tables.
#20 Solving Radical Equations	#10, #15	Graphically solving equations containing radicals

INTRODUCTION OF KEYS

Unit Title	Keys	
10: Graphical Solutions to Linear Equations		**Y =** **WINDOW** **TRACE** **GRAPH** **QUIT**
	CALC	**5:intersect**
11: Linear Applications	**ZOOM**	**6:ZStandard** **3:Zoom Out** **TblSet** **TABLE**
12: Graphical Solutions to Linear Inequalities	**Y-vars**	**1:Function** **1:Y1** **TEST menu**
13: Graphical Solutions to Absolute Value Equations		**No new keys**
14: Graphical Solutions to Absolute Value Inequalities		**No new keys**
15: Solving Factorable Polynomial Equations	**ZOOM** **CALC**	**2:Zoom In** **2:root**
16: Solving Non-factorable Quadratics		**No new keys**
17: Applications of Quadratic Equations	**CALC**	**4:maximum** **1:value (EVAL X)**
18: Solving Quadratic Inequalities		**WINDOW ▸ FORMAT** **AxesOn AxesOff**
19: Solving Nonlinear Inequalities		**MODE** **Connected Dot**
20: Solving Radical Equations		**No new keys**

An equation is a symbolic statement that two algebraic expressions are equal. Solving an equation means you find a replacement value for X that will produce the same value for both expressions. This unit will consider first degree conditional equations in which there will be one correct solution, identities, for which there are infinitely many solutions and contradictions for which there are no solutions.

1.   Solve 5X - 1 = -3, algebraically: Check your solution by

      substitution:

2.   To graphically solve this same equation we will look at a graphical representation of the algebraic expressions on each side of the equation. Press [Y=] and [CLEAR] to delete any expressions. The left side of the equation, 5X - 1, will be designated as Y1. Enter 5X - 1 after "Y1 = " (remember to use the [X,T,θ] key). The right side of the equation, -3, will be designated as Y2. Enter -3 after "Y2 = " (be sure to use the gray (-) sign). We want to determine graphically where Y1 = Y2.

3.   Press [WINDOW] and cursor down to each line to enter the following values:

      These values designate the size of the viewing window when we graph Y1 and Y2. These WINDOW values were designed to give you "nice, friendly" numbers when you use the TRACE feature in #5.

```
WINDOW FORMAT
 Xmin=-9.4
 Xmax=9.4
 Xscl=1
 Ymin=-6.2
 Ymax=6.2
 Yscl=1
```

4. Press [GRAPH]. The graphical solution to the equation is the intersection of Y1 and Y2. At the intersection point, Y1 is equal to Y2. Circle the intersection point.

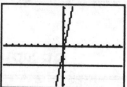

5. We want the **X value** that is the solution to the equation. The X value can be found by pressing [TRACE] and tracing along the graph to the point of intersection. (The left and right arrow keys will move the cursor along the path of the graph. The small "1" or "2" in the upper right corner will tell whether you are on the graph of Y1 or Y2.) At the intersection point, X = _____. At this value the equation 5X - 1 = -3 evaluates to a true arithmetic sentence. (Does this agree with the solution you got in #1?)

6. How can you be sure that your cursor was on the <u>exact</u> point of intersection? When you think you have reached the point of intersection, record the values indicated at the bottom of the screen. _____ Press the up (or down) arrow key once to "jump" to the other graph. You will know you are on a different graph because the number in the upper right corner of the screen will change. Compare the numbers that are now displayed on the bottom of the screen to those you recorded. If they are <u>both</u> the same, then you are on the exact point of intersection. If not, adjust your TRACE cursor and test again.

7. Solve 3 - 2X = 2X + 7 graphically. Press [Y=] and clear the expressions after Y1 and Y2. Enter 3 - 2X after "Y1=" and 2X + 7 after "Y2=". Press [TRACE] and trace along the line to the point of intersection. The solution is X = _____. (Be sure that you are on the exact point of intersection.)

8. Solve 3X + 4 = 2 - X graphically. Press [Y=] and clear the expressions after Y1 and Y2. Enter 3X + 4 after "Y1=" and 2 - X after "Y2=". If you try to TRACE to the point of intersection you quickly discover that the calculator will not allow you to reach the <u>exact</u> point of intersection. We will use the calculator's INTERSECT option to find the point of intersection.

The INTERSECT option will be used exclusively from this point on. It is not dependent on the WINDOW values you entered in #3. The only requirement is that you be able to <u>see</u> the point of intersection. Set your WINDOW values to those displayed on the screen at the right. Press [WINDOW] and use the down arrow key to move down the screen and change the values. We are doing this to enlarge the viewing window.

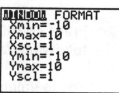

9. To access the INTERSECT option, press [2nd] <CALC>. ("CALC" is located above TRACE.) Press [5] to select INTERSECT. Move the cursor along the first curve to the approximate point of intersection and press enter. At the "second curve" prompt you can press [ENTER] again because the cursor will

90

still be close to the point of intersection. At the "guess" prompt, press **[ENTER]**. (We are telling the calculator that our "guess" is the approximate point of intersection that we designated at the "first curve" prompt.) The solution for X is ____. Your screen should look like the one at the right.

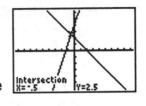

10. Now solve $3X + 4 = 2 - X$ algebraically.

Is your answer equivalent to the answer you got in #9? Return to the Home Screen (**[2nd] <QUIT>**). The value you got for "X" in #9 is now stored in X. You can retrieve this value by entering X (press **[X,T,θ]**) and then convert the value of X to a fraction, if necessary, by pressing **[MATH] [1:▶Frac] [ENTER]**.

11. Now check your solution. Refer back to Unit #8, step 7 if you do not remember how to do this. Compare the results of your "check work" to the information displayed on the INTERSECT screen shown in #9. What correspondence do you see? What does the X value represent on the graphing screen and what does the Y value represent?

EXERCISE SET

Using the INTERSECT option, solve these equations graphically. Follow the same procedure as is outlined in #9 above. Sketch the INTERSECT screen that yields the solution and use ▶Frac to convert all decimal answers to fractions.

A. $5 = 2 - 7X$

X = _____

Converted to a fraction, X = ____

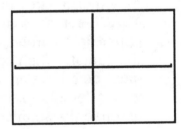

B. $(-4/3)X = -2$

X = _____

Converted to a fraction, X = ____
*(In your textbook, this problem would look like

this: $-\dfrac{4}{3}x = -2$)

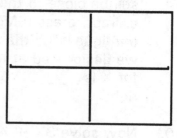

C. $\dfrac{-4x}{3} + 6 = -1$

X = _____

Converted to a fraction, X = ____

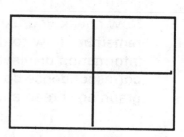

D. $\dfrac{2x - 1.2}{0.6} = \dfrac{4x + 3}{-1.2}$

X = _____

Converted to a fraction, X = ____

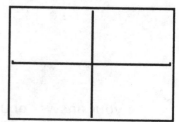

Did you get -3/40 (i.e. -.075)? If you did not, check the way you entered the equation. You will need to insert parentheses in the appropriate places.

12.   Solve $4(X - 1) = 4X - 4$ graphically. Press **[Y=]** and
        enter 4(X - 1) after "Y1 =" and 4X - 4 after "Y2 =".
        Press **[TRACE]**. ONLY ONE graph is displayed!!
        Trace along this line and observe the number in the
        upper right corner of the screen. Which graph are
        you tracing on?_____ Now use the up (or
        down) arrow key (press only one) and move the
        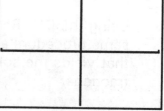
        TRACE cursor to the other graph. Again check the number displayed in the
        upper right corner, which graph are you on now?_____ Both graphs are
        the same! When both graphs are the same, then both sides of the equation
        must be equivalent expressions. Equivalent expressions produce identical
        values for all replacement values of the variable. This means that you have
        graphed the solution to an IDENTITY. Identities are true for all values of X
        that are acceptable replacement values for the variable in the equation. Thus
        the solution to this equation is the set of all real numbers.

13. Solve 2X - 5 = 2(X + 1) graphically. Press [Y=] enter 2X - 5 after "Y1" and 2(X + 1) after "Y2=". Press [TRACE]. Observe that the two lines are parallel. Parallel lines never intersect and hence there is no solution. We indicate that this equation has no solution by writing the symbol for the empty set. This equation is called a contradiction.

EXERCISE SET CONTINUED

Solve the equations below graphically. Sketch the screen that is displayed and identify the type of equation.

E.   3(X - 1) - 2 = 4X - 7 - X - 2

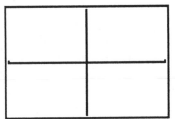

What type of equation is this? (identity or contradiction)

_____

F.   2(2X - 5) + 7 = 4X - 3

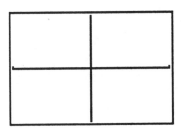

What type of equation is this? (identity or contradiction)

_____

NOTE: *The equations in this unit were specifically written to conform to the WINDOW values displayed on the screen at the right. If you plan to use your calculator to solve equations in your textbook, you may not be able to see the intersection point displayed on the screen. You can remedy this problem by adjusting the WINDOW values. Begin by increasing the Xmax and Ymax by 5 and decreasing the Xmin and Ymin by 5 units. (i.e. [-15,15] by [-15,15]) Continue to increase/decrease by increments of 5 until the points of intersection are displayed.*

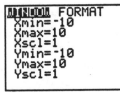

14. Summarizing Results: Write a summary of what you learned in this unit. You should address the following:
    a. the manner in which you enter equations on the TI-82,
    b. use of the INTERSECT option,
    c. graphical representation of the three types of linear equations (conditionals, identities, and contradictions).

15. Highlight #5, #6 and #9 of this unit for your QUICK REFERENCE list of operations.

<u>Solutions to Exercise Sets:</u> **A.** -.4285714, $-\dfrac{3}{7}$ **B.** 1.5, $\dfrac{3}{2}$ **C.** 5.25, $\dfrac{21}{4}$ **D.** -.075, $-\dfrac{3}{40}$

E. contradiction **F.** identity

There are substantial differences in the commands for graphing between the TI-82 and the TI-85. Refer to the TI-85 GUIDEBOOK for a full explanation of graphing with the TI-85. Here we will explain only those differences which relate to the problems and concepts in Unit #10.

To graph using the TI-85, first select [GRAPH]. This will bring up a menu at the bottom of the screen. These options correspond to the five buttons at the top of the TI-82 calculator, with RANGE corresponding to WINDOW.

2.   Select [F1](y(x) =) and enter the two functions as in step #2. Pressing [ENTER] after typing in y1 will automatically display "y2 = ". Only functions whose equal signs are highlighted will be graphed. To select a function, press [F5](SELECT).

3.   Select [F1](RANGE), and for a "friendly" window on the TI-85 enter xMin = -12.6, xMax = 12.6, with the other values the same as on the first screen in Unit #10.

4.   In the GRAPH menu, press [F4](TRACE) to activate the TRACE feature..

5.   The TI-85 has an INTERSECT option, as does the TI-82. However, its method of access is different. With the [GRAPH] menu displayed, and the graphs of the functions showing, press [MORE] to see more options on the menu. The left-most option is MATH. Press [F1](MATH) to choose this option, then [MORE], and choose the ISECT option by pressing [F5]. (ISECT stands for InterSECTion). With your cursor on one of the graphs near the desired intersection press [ENTER]. The cursor will then "jump" to the other graph. You should again move the cursor close to the point of intersection and press [ENTER] again. Your screen will display the word "ISECT" and the x and y coordinates of the intersection. If more than one intersection of the two graphs is on the screen, you must use the ISECT function for each of them.

UNIT #10 APPENDIX
Casio fx7700GE

There are substantial differences in the commands for graphing between the TI-82 and the Casio. Refer to the *fx-7700GE Owner's manual* for a full explanation of graphing with the Casio. Here we will explain only those differences which relate to the problems and concepts in Unit #10.

To graph using the Casio, first select [GRAPH] from the MAIN MENU. This will bring up a list consisting of Y1, Y2, ..., Y5, then a space with a blinking cursor. In this space, type the first function, 5X-1, then push [F1](STO), select Y1 and push [F6](SET). Type [(-)] [3] [F1] (ST0). Then arrow down to Y2 and press [F6](SET).

Select [Range], and for a "friendly" window on the Casio enter the same values as for the TI-82, then press [EXIT] to return to the function screen.

Press [F6] for [DRW], then [F1] for [TRACE] as you would on the TI-82.

The Casio does not have an INTERSECT option.

2.   To mimic the operations of the TI-82, save each expression as variable using the VAR menu. For example, enter the expression 5X - 1 onto the stack by pressing ['] [5] [×] [α] <X> [-] [1] [ENTER] then press [Y] [1] [STO]. This defines a variable Y1 with the expression 5X - 1. In a similar manner, define the variable Y2 to be -3.

3.   To graph an expression or equation, enter the PLOT application by pressing [→] <PLOT>. The TYPE should be set to Function. Move the cursor to the EQ line. Press (CHOOS), then select your variables by moving the cursor and pressing (✓CHK) at each variable. (Note that you can directly enter an expression to be graphed, or you can enter a list of expressions rather than defining a variable. A list is surrounded by { } which is accessed by pressing [←] [{ }]. Each expression should be surrounded by single quotes and separated by a space which is done by pressing [SPC].) The independent variable INDEP should be set as X. In order to get "nice, friendly" numbers when you use the TRACE feature set the minimum horizontal value to - 6.5, the maximum horizontal value to 6.5, the minimum vertical value to - 6.2, and the maximum vertical value to 6.4. Your screen should look like the one at the right when you are done.

```
░░░░░░░░░░░░PLOT░░░░░░░░░░░░
TYPE: Function      ∡: Deg
EQ:   { -3 '5*X-1' }
INDEP: X   H-VIEW: -6.5 6.5
_AUTOSCALE V-VIEW: -6.2 6.4
ENTER FUNCTION(S) TO PLOT
 EDIT CHOOS      OPTS ERASE DRAW
```

The Xscl and Yscl option in the TI-82 sets the number of units between tick marks. In order to set the interval of each vertical and horizontal tick mark on the HP 48G you must select OPTS menu key from the main PLOT screen. Then you must set X-TICK and Y-TICK to number of units you want for each tick. In this case set each to 1. Then you want the check mark next to PIXELS on that same line to be turned off. To do this, move the cursor on the check mark and press [+/-]. This toggles the check on and off. We must do this because the default setting on the HP 48G is to have each tick mark 10 pixels apart. (You do not need to concern yourself if you do not know what a pixel is. This will be covered in Unit#21.) The PLOT OPTIONS screen should look like the one at the right.

```
░░░░░░PLOT OPTIONS░░░░░░
INDEP: X   LO: Dflt   HI: Dflt
∡AXES   _CONNECT   _SIMULT
STEP: Dflt _PIXELS
H-TICK: 1   V-TICK: 1   _PIXELS
ENTER INDEPENDENT VAR NAME
 EDIT            CANCL OK
```

Press (OK) to return to the PLOT window.

4. Press (DRAW) to plot your expressions. This is called the PICTURE environment. You should always press (ERASE) before plotting to clear any old plots or else the new plot will be plotted with the old.

5. Press (TRACE) to activate the TRACE mode. Then press ((X,Y)) to display the coordinates of the current cursor position. As in the TI-82, the left and right arrow keys are used to move the cursor. Pressing ((X,Y)) hides the menu; to redisplay the menu, press [NXT]. Pressing (CANCL) or [ON] returns you to the main PLOT screen.

9. The ISECT menu command is equivalent to the INTERSECT option for the TI-82. To access the ISECT command press (FCN) in the PICTURE environment. Then press (ISECT). The point of intersection will be displayed at the bottom of the screen.

    To return the menu back to the main PICTURE environment menu, press [→] <MENU>.

UNIT #11
LINEAR APPLICATIONS

For each problem below, identify your variable and what it represents (use a "Let" statement). Write an equation to represent the problem. We will experiment with using the calculator for solving.

1.   A jogger can average a speed of 8 mph on short runs. To run a distance of 6 miles, how long will it take him? (Remember, distance = rate • time).

   a. Let X = _____

   b. Equation:_____

   Press **[ZOOM] [6]** (this sets up a standard viewing screen). Press **[Y=]** and enter the left side of the equation at the **Y1=** prompt and the right side of the equation at the **Y2=** prompt (this is the same technique that you used in the previous unit). Press **[GRAPH]** to see a picture of the relationship. Recall that the X-value of the point of intersection is the solution to your original equation; it is where Y1=Y2. Have the calculator find the solution for you by using the **INTERSECT** option you learned in Unit #10 (you may want to go back and review the unit, beginning at step 9).

   c. Write the answer to the above problem in a sentence below.

2.   A length of a rectangle is twice the width. If the perimeter is 60 inches, what are the dimensions of the rectangle?

   a. Let X = _____

      2X = _____

   b. Recall, Perimeter = 2(length) + 2(width)

      Equation: _____

Enter the left side of your equation at Y1, the right side of the equation at Y2. Press **[GRAPH]**.

Because you only see one line (and no point of intersection), the calculator cannot perform the calculation. Your screen should look like the one at the right.

To make the calculator "back up" and give you a better picture, we will ZOOM OUT. Before proceeding, press **[ZOOM] [▶] [4:SetFactors]**. Both the XFact and the YFact should equal 4. If not, cursor down and enter 4. This zoom factor of 4 means that the WINDOW values will increase by a multiple of 4 each time you ZOOM OUT.

Press **[ZOOM] [3]** (accessing the "zoom out" option) **[ENTER]**. Since you still see only one line, press **[ENTER]** again (activating the "Zoom out" option again). You should now be able to see the point of intersection (see the screen display at the above right). Use the INTERSECT option of CALCULATE to find the X-coordinate of the point of intersection. Once you have determined that value, answer the question posed in the original problem:

The width of the rectangle is _____ inches.

The length of the rectangle is _____ inches.

NOTE: The axes thickened because the X and Y scales on the WINDOW screen were not set at zero. You may reset these scales to 0 any time you use the ZOOM OUT option.

3. We know that money invested in a savings account at a given rate earns interest. The formula we can use to find the amount of money that would be in an account after X number of years (assuming a constant interest rate per year) is

Amount = principal + (principal)(rate)(time)

Assume you are interested in the time it would take an investment to double when earning 5% interest annually. Write an equation that represents the amount of time it would take for an investment of $50 to double to $100. Let X = the number of years required.

a. Equation:_____

Enter the left side of the equation as Y1 and the right side of the equation as Y2. Press [ZOOM] [6].* What do you see?

* NOTE: Since we ZOOMED OUT twice on the last problem we must reset our viewing window to ZStandard by pressing [ZOOM] [6].
We need to see a bigger picture. To get one, ZOOM OUT as you did in #2. Record your calculator screen at the right, circling the point of intersection.

Use the intersect feature to calculate the intersection of the two lines (and our desired X value).

X = _____

Answer the question posed above, how long will it take your investment of $50 to double to $100 when invested in an account earning 5% annually?

4. What if we invested $500 initially rather than $50? Would it still take 20 years for the investment to double?

    Look at the expressions you entered for the previous problem at Y1 and Y2. Use the <INS> feature of the calculator to insert zeroes appropriately, then press [GRAPH] to see a picture of the relationship. Hint: you may have to Zoom Out some more before you see the intersection of the two lines. How many **more** times do you need to ZOOM OUT? _____ Do you have a clear picture of the relationships? Why or why not?

    Can you still use the Intersect feature to calculate the point of intersection? Try it and see.

    Conclusion: Would it still take 20 years to double the investment?

5. For the problem below, clearly identify your variable and write an equation representing the problem.

    In 1991 a state-of-the-art computer system cost $3,000. It depreciates at an average rate of $200 per year. After how many years will the system be worth $1600?

a. Let X = _____

b. Equation:_____

Using the techniques above, we enter the left side of the equation at Y1 and the right side of the equation at Y2. We begin with a standard viewing screen ([ZOOM] [6]).

What do you see initially?

To get a better picture (actually, just a picture!) we can access our zoom option. Press [ZOOM] [3] [ENTER]. Notice the thickening of the axes, and the appearance of one line. However, we entered two expressions, and should be seeing two distinct lines. Moreover, to have the calculator compute the intersection (the solution) of the equation we must be able to actually see the intersection.
Press [ENTER] again to "ZOOM OUT".

Now what do you see?

Press [ENTER] again, describe what you see.

Press [ENTER] again, and describe what you see.

Obviously, your picture is not clear, you are frustrated, and have probably forgotten what you even wanted to know!

> There must be a better way! You should not spend
> more time trying to graphically solve the equation
> than you would if you were solving "by hand".

At this point, we will explore the options that are available for use.

6. Recall your equation, 3000 - 200X = 1600. The variable "X" represents the number of years it would take your state-of-the-art computer system (whose initial cost was $3000) to depreciate to $1600.

The TI-82 has a TABLES feature. To access it, we must first set our table by pressing [2nd] <TblSet>. You should see a screen like that at the right. "TblMin" refers to the minimum value you want set for your table. Because we are referring to years, we will begin our table with the number "1". Cursor down to "ΔTbl".

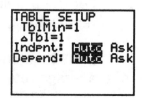

"ΔTbl" refers to the manner in which you want your table incremented. In other words, do you want the X-values to increase by 1, 2, 3, etc.? We will increment our table by ones since we are looking for a specific number of years. The next two settings should have "Auto" highlighted since we want the calculator to automatically compute both the independent and dependent variables for us. For each entry in the X-column, the calculator evaluates the expression entered at Y1 = , Y2 = , etc. and displays the value returned for each expression.

To actually see the table press [2nd] <Table>. The given screen will be displayed.

X	Y₁	Y₂
1	2800	1600
2	2600	1600
3	2400	1600
4	2200	1600
5	2000	1600
6	1800	1600
7	1600	1600

X=1

Notice that our X-values begin with 1 and are incremented sequentially by ones. The calculator has computed the corresponding Y value for both the equations entered at the Y1 prompt and the Y2 prompt (remember, our Y2 was just the number 1600).

For what X-value are the two Y-values equal? _____ (This is the point where our two lines would have intersected **if** we could have displayed a "nice" picture.)

What does this mean in terms of our problem? (write a complete sentence):

7. Scroll down the X-column of the table, using the down arrow key. After 10 years, what is the expected value of your computer system?_____

What could you expect to get for your computer after 15 years?_____

8. In leasing a copy machine for the university, 2 plans were proposed. From the KOPY-IT machine rental company, the cost would be $200 a month for machine usage, with a charge of $15 added for each ream of paper used. The DUPLICATE CO. offered a rental fee of $300 a month, with a charge of $5 per ream of paper. Find the break point. (The break point is the number of reams of paper used to make the cost equal, regardless of which company is contracted.)

Let X = the number of reams used

a. Write an expression for the monthly cost for KOPY-IT: _____

Write an expression for the monthly cost for DUPLICATE CO.:_____
   Enter your expression for the KOPY-IT company at Y1, and your
   expression for the DUPLICATE CO. at Y2.

We want to use the calculator to find the break point. This will be computed
in two ways, by using the INTERSECT option of the calculator and by
accessing the TABLE feature. Record your investigation step by step so that
someone else could follow your steps and get the same result. Record your
findings in the space below for each method.

INTERSECT option:
(Start this process by pressing [ZOOM] [6] to set the
standard viewing window.)

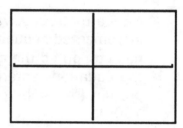

If you had to ZOOM OUT, record the number of times you zoomed._____

TABLE feature:

ΔTbl = _____

Y1 = _____

Y2 = _____

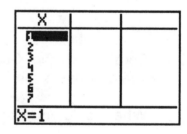

Record values displayed in TABLE for Y1 and Y2.

9.   You have experimented with using the calculator to solve linear equations in
     one variable using both the INTERSECT option and the TABLE feature. You
     need to remember that the calculator should enhance your problem-solving
     capabilities - NOT frustrate you. The only way to reach this point is to
     practice, and record both what you want the calculator to do for you and
     HOW YOU are directing the calculator to do it! Practice doing problems in
     your textbook using the techniques you have discovered in this unit.

10. Summarizing Results: Summarize what you learned in this unit. Your summary should address the following points:

    a. Use of the INTERSECT option of the calculator.
    b. Use of the TABLE feature (addressing TABLE SETUP in your discussion).

11. Highlight #2 and #6 of this unit for your QUICK REFERENCE list of operations.

NOTE: While experimenting with the calculator when working your textbook applications, keep a log of what you discover. Some of you will like "Zooming" and then using the CALCULATE feature to find the desired intersection. Others, will prefer the TABLE feature. Still others will prefer paper and pencil solutions. Remember, the calculator gives a picture of relationships and is a tool to help make your work easier. Play with it and experiment to find how it best serves YOUR needs.

<u>Solutions:</u> **1a.** time, **1b.** $8X = 6$, **1c.** It would take him .75 hrs. to run 6 miles. **2a.** $X = $ width, $2X = $ length, **2b.** $60 = 2(2X) + 2X$, 10, 20, **3a.** $100 = 50 + 50(.05)(X)$, you see nothing, $X = 20$, It will take 20 yrs. for the investment to double., **4.** 2 ZOOM OUTS, No, because one line is so vertical that it appears to be a part of the Y-axis., yes, yes, **5a.** years, **5b.** $3000 - 200X = 1600$, You initially see nothing, but eventually will see a thickening of the Y-axis and a horizontal line., **6.** $X = 7$, In 7 yrs. our computer will have a value of $1600., **7.** $1000, $0 **8a.** $200 + 15X$, $300 + 5X$, 3 ZOOM OUTS, First TABLE entry line should be $X = 1$, 215, 305.

To access the **ZOOM** features of the TI-85, you must first press **[GRAPH]** then **[F3](ZOOM)**. (If you have two rows of menu displayed you will need to press **[2nd]** **<M3>(ZOOM)**.) Now you can move back and forth using **[MORE]** and the menu keys to select your choices. For example, to get a standard viewing window, select **[F3](ZOOM) [F4](ZSTD)**. Similarly, ZOOM OUT is **ZOUT**, ZOOM BOX is **BOX**, and **ZPREV** will return you to the previous viewing window.

To check the zoom factors, as indicated in section 5, press **[ZOOM] [MORE]** **[MORE] [ZFACT]** and set the xFact to 4 and the yFact to 4.

Regarding section 6 and on: The TI-85 *does not* have a TABLES feature.

```
┌──────────────────────────────────────────────┐
│            UNIT #11  APPENDIX                  │
│            Casio fx7700GE                      │
└──────────────────────────────────────────────┘
```

The **ZOOM** features of the Casio are quite similar to those of the TI-82, with the
distinct advantage of the [ZOOM] [ORG] option, which returns you to your original
window. For example, to zoom in on a certain area of the graph, select [x f] from
the ZOOM menu. Similarly, ZOOM OUT is [x 1/f].

Tables

Regarding section 6 and on: The Casio *does not* have a TABLES feature.

UNIT #11 APPENDIX
HP 48G

The HP 48G does not have a standard viewing screen (ZStandard for the TI-82). You must manually set the horizontal minimum and maximum to -10 and 10 and the vertical minimum and maximum to -10 to 10 in the main PLOT screen, and make sure that the horizontal and vertical ticks are 1 and that PIXELS is not checked. The HP does have a default setting (ZDFLT). This will set the horziontal range from -6.5 to 6.5 and the vertical range from -3.1 to 3.2 and in PLOT OPTIONS set horizontal and vertical ticks to 10 pixels.

To "zoom out", select ZOOM menu key while in the PICTURE environment then press (ZOUT).

The HP 48G *does not* have a TABLES feature.

1. To solve linear inequalities such as 5X - 1 $\geq$ -3, we want to find replacement values for X that will produce a true arithmetic sentence. The solution to a first degree inequality in one variable is typically an infinite set of numbers, rather than a single number.

    Solve 5X - 1 $\geq$ -3, algebraically:

2. The -2/5 you got in your solution is a "critical point". It is so named because it divides the number line into three distinct subsets of numbers - those larger than the number, those smaller than the number, and the number itself.

    a. In the space below, test the critical point -2/5 in the original inequality. (i.e. When X is replaced by -2/5, is the resulting inequality a true statement?)

    b. In the space below, test a number whose value is larger than that of your critical point.

    Because your test number tested true, this implies that all values to the right of the critical point will also test true. (This is also indicated by the mathematical statement x $\geq$ -2/5, your algebraic solution found in #1.)

    Number Line Graph of your solution: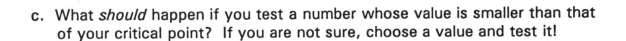

    c. What *should* happen if you test a number whose value is smaller than that of your critical point? If you are not sure, choose a value and test it!

109

3. The graphics calculator can be used to quickly test your solution of $x \geq -0.4$. Press [Y=] and enter 5X - 1 after "Y1=". Return to the home screen by pressing [2nd] <QUIT>. Now select a value for X that is greater than -0.4, for example, 3. Store 3 in X, enter a colon and evaluate Y1. Y1 can be located by pressing [2nd] <Y-vars> [1] [1] [ENTER]. Your display should look like the one at the right. We could have evaluated 5X - 1 at any value of X greater than -0.4. Our result should be greater than -3 (which is the right side of the inequality) if our solution is correct.

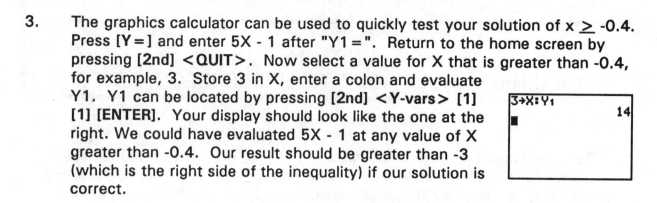

4. Now consider the graphical solution. Press [Y=] and after "Y2=" enter the expression -3. Since 5X - 1 = Y1 and -3 = Y2 , we want to find graphically where Y1 ≥ Y2.

5. Press [ZOOM] [6] to set the standard viewing WINDOW. The WINDOW values at the right are automatically entered and the graph screen will be displayed. (Pressing [WINDOW] allows you to see these values.)

6. Press [GRAPH] to view the graphical solution to the inequality. Sketch the picture displayed and circle the point where the 2 graphs are equal. (i.e. point of intersection)

7. Use the INTERSECT option to determine the solution to the equation. You should get X = -0.4. Press [TRACE] and TRACE along Y1. (You will know you are tracing along Y1 by the "1" in the upper right corner.) Label this line with "Y1" and the other line with "Y2". Graphs should be read from left to right. As you TRACE, the graph is a visual of changes in Y-values as the X-values increase. Which section of Y1 is greater than Y2? It should be the section that is highlighted on the graph at the right. In general, Y1 > Y2 where Y1 is above Y2 on the graph.

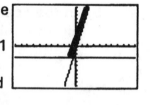

8. TRACING from left to right along Y1, you should observe the following:
   i. The X-values increase as you TRACE from left to right.
   ii. At the point of intersection, Y1 = Y2.
   iii. As you move to the right of the intersection point, the Y1 values become larger than Y2 (remember, Y2 = -3).
   The solution to the inequality is X ≥ -0.4.
   In set notation this would be written: $\{X | X \geq -0.4\}$
   As a number line graph this would be:

   -.04

   In interval notation this would be written
   $[-0.4, \infty)$.

A.   **Solve** 10 - 3X < 2X + 5 graphically.

Solution Steps:
i. Press [Y=] and enter 10 - 3X after "Y1="
and 2X + 5 after "Y2=". Sketch the graph
displayed. Be observant the first time the
graphs are displayed. It will be helpful to label Y1, which is the first
graph to be displayed, on your sketch.

ii. Use the INTERSECT feature to find the point of intersection. The
intersection is at X = _____. The expressions entered at Y1 and Y2
are equivalent when X = 1. Therefore, X = 1 would be the solution to
the _equation_ 10 - 3X = 2X + 5.

iii. [TRACE] along the portion of Y1 that is below the graph of Y2. As
you TRACE you should observe that the further below Y2 you go, the
larger the X values become. Use a highlighter pen to highlight the
section of Y1 that is **less than** Y2. (Y1 < Y2 when the graph of Y1 is
below the graph of Y2.)

iv. Conclusion: Y1 < Y2 when X > 1.
The solution set will be { X |X > 1}.

Translate this solution to a number line graph:

B.   **Solve** 6 - 5X ≤ -1 graphically, following the
steps outlined in A above.

Solution Set:_____

Translate this solution to a number line graph:

111

C.   **Solve -3 ≥ 7 - 2X graphically, following the** steps outlined in A.

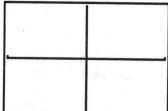

Solution Set:_____

Translate this solution to a number line graph:

D.   Solve $\dfrac{4X - 2}{6} < \dfrac{2(4 - X)}{3}$ graphically,

following the steps outlined in A.

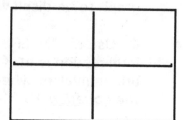

Solution Set:_____

Translate this solution to a number line graph:

TEST MENU

You may use the calculator's TEST menu to display a graph that resembles the number line graph of any of the inequalities you have solved. This provides a means of counterchecking your work. For the example, we will check the original inequality, 5X - 1 ≥ -3 with the TEST feature.

9.   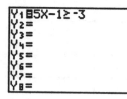   The calculator will display a graph that resembles the number line graph of your solution set. TEST is located above the MATH key. Begin by deleting all entries on the "Y =" screen. Enter the entire inequality, 5X - 1 ≥ -3, on the **Y1=** line. Press **[GRAPH]**.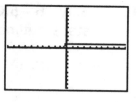

NOTE: If you TRACE on this graph, the TEST will return a 1 as the Y-value for a true statement and a 0 as the Y-value for a false statement.

10. Compare this number line graph to the one you sketched in #2. **BEWARE!** If you TRACE on this number line you will not be able to determine the critical point of your number line. This point must still be determined using the procedure outlined in Exercise A (graphing each side separately and using the INTERSECT option). However, the graph illustrated in #9 <u>does</u> show you whether the interval to the right or the interval to the left of the critical point is the solution interval.

EXERCISE SET CONTINUED

Directions: Enter the inequality at the Y1 prompt. Press [GRAPH] and sketch the display. Transfer the information displayed to your number line and then convert the information to interval notation. You will need to determine the critical point on your number line by referring back to the corresponding problem in Exercises A through D. Be sure your calculator display agrees with the number line you sketched in each of the Exercises A - D.

E. Use the TEST menu to display a graph that resembles the number line graph of the solution to the inequality $10-3X < 2X+5$.

Number line graph: ⟵—————————⟶

Interval Notation:_____

F. Use the TEST menu to display a graph that resembles the number line graph of the solution to the inequality $6 - 5X \leq -1$. Be sure your calculator display agrees with the number line you sketched in Exercise B.

Number line graph: ⟵—————————⟶

Interval Notation:_____

G. Use the TEST menu to display a graph that resembles the number line graph of the solution to the inequality $-3 \geq 7-2X$. Be sure your calculator display agrees with the number line you sketched in Exercise C.

Number line graph: ⟵—————————⟶

Interval Notation:_____

113

H. Use the TEST menu to display a graph that resembles the number line graph of the solution to the inequality

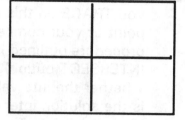

$$\frac{4X - 2}{6} < \frac{2(4 - X)}{3}$$. Be sure your

calculator display agrees with the number line you sketched in Exercise D.

Number line graph: ←——————————→

Interval Notation: _____

11. In your own words, explain why you cannot use the graph that is displayed on the graph screen in Exercises E - H as your only means of solving an inequality. (i.e. What is the one **major** obstacle that this approach has?)

NOTE: *The inequalities in this unit were specifically written to conform to the WINDOW values displayed on the screen at the right. If you plan to use your calculator to solve inequalities in your textbook, you may not be able to see the intersection point displayed on the screen. You can remedy this problem by adjusting the WINDOW values. Begin by increasing the Xmax and Ymax by 5 and decreasing the Xmin and Ymin by 5 units. (i.e. [-15,15] by [-15,15]) Continue to increase/decrease by increments of 5 until the points of intersection are displayed.*

12. Summarize Results: On the next page, summarize what you have learned in this unit about graphically solving linear inequalities. Be sure to include:
   a. a comparison of solving an <u>equation</u> vs. solving an <u>inequality</u>
   b. how to determine the critical point and whether or not it is included in the solution, and
   c. how to decide whether to include the numbers greater than the critical point in your solution set or those numbers that are less than the critical point.

13. Highlight Exercise A, solution steps, of this unit for your QUICK REFERENCE list of operations. You may also want to highlight #9 and #10 if you like using the TEST menu.

Solutions to Exercise Sets: A. $\overleftarrow{\quad\underset{1}{\oplus}\longrightarrow}$ B. $\{X|X\geq1.4\}$ $\overleftarrow{\quad\underset{1.4}{\bullet}\longrightarrow}$ C. $\{X|X\geq5\}$ $\overleftarrow{\quad\underset{5}{\bullet}\longrightarrow}$

D. $\{X|X<2.25\}$ $\overleftarrow{\quad\underset{2.25}{\oplus}\longrightarrow}$ E. $\overleftarrow{\quad\underset{1}{\oplus}\longrightarrow}$ $(1,\infty)$ F. $\overleftarrow{\quad\underset{1.4}{\bullet}\longrightarrow}$ $[1.4,\infty)$ G. $\overleftarrow{\quad\underset{5}{\bullet}\longrightarrow}$ $[5,\infty)$

H. $\overleftarrow{\quad\underset{2.25}{\ominus}\longrightarrow}$ $(-\infty,2.25)$

115

The TI-85 does have a TEST menu. It is accessed using [2nd] <TEST> (located above [2]). Your selection is made from the menu at the bottom of the screen.

Number Lines

The Casio cannot be used to give number line solutions of linear inequalities in the same way as the TI-82 does in problems 9 and 10.

3. You should define a variable Y1 to be the expression 5X-1 and the variable X to be -3. Then select the variable Y1 from the VAR menu and evaluate it using the EVAL key.

TEST MENU

The HP 48G *does not* have a feature similar to the TI-82's TEST feature.

Absolute value can be used to measure the distance between two points on a number line. The distance between the points can be expressed as the absolute value of the difference between the two coordinates. For example, the distance between 3 and -3 would be written as $|3 - (-3)| = |3 + 3| = |6| = 6$. Absolute value always yields 0 or a positive number. Thus, absolute value must be used to indicate distance because measurement must be in terms of non-negative numbers.

1. What would be the distance between -9 and -3?
    Solution: Because distance is the absolute value of the difference between coordinates:

    $$d = |-9 - (-3)| = |-9 + 3| = |-6| = 6$$

    The distance between -9 and -3 is 6 units on the number line. You will notice that distance between 3 and -3 in the above example was also 6.

2. The two above problems could be generalized in the following ways:
    a.   If the distance between a number and -3 is 6 units, find the number.

    OR

    b.   What numbers "X" are 6 units from -3?
         Solution: $|X - (-3)| = 6$, which is the same as $|X + 3| = 6$.
         Use the graphing calculator to solve this equation graphically by graphing each side of the equation and find the point(s) of intersection. The X value at the point(s) of intersection will be the solution(s) to the equation. Enter $|X + 3|$ after "Y1=" on the "Y=" screen by pressing [Y=] and entering **abs(X + 3)** after "Y1=". Remember, $|X + 3|$ is read: the absolute value of the quantity X plus 3. Now enter 6 after "Y2=".

3. Press **[ZOOM] [6]** to automatically enter the values on the WINDOW screen at the right. Even though the keystrokes take you straight to the graphical display, you can confirm these WINDOW values by pressing **[WINDOW]**. Do this now. Your screen should look like the one at the right.

```
WINDOW FORMAT
Xmin=-10
Xmax=10
Xscl=1
Ymin=-10
Ymax=10
Yscl=1
```

119

Now press **[GRAPH]** to display the graphical representation of the absolute value equation. Your screen should look like the one at the right.

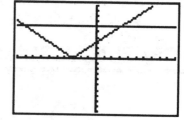

a. Circle the two points of intersection that are displayed.

b. Use your ruler and draw a dotted vertical line from the points of intersection perpendicular to the horizontal X-axis.

c. Count the tic marks on the axis and label the value of each of these points.

d. Count the tic marks and label the location of the point with the value of -3.

e. How many units are between -3 and 3?_____

f. How many units are between -3 and -9?_____

4. Now use the calculator's INTERSECT option to find the X values at the points of intersection. See UNIT #10 if you need a refresher on how to use the INTERSECT option. You will need to use the INTERSECT option <u>twice</u> since there are two points of intersection. Copy the screen display to show where you found <u>one</u> of the two solutions, but record both of the solutions here:
X = _____ or X = _____

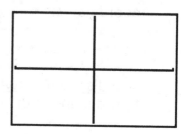

NOTE CONCERNING THE USE OF THE INTERSECT OPTION: In the future there may be other equations that have more than one solution. There will be an intersection point for each of these solutions that is a real number. The INTERSECT option will have to be completed for each point of intersection. To justify your work you will only be required to sketch one of the INTERSECT screens that you used and merely record the solutions derived from the other screens.

5. THINK ABOUT IT! Could you use just the number line (horizontal axis) and solve this equation graphically **without** using the graphing calculator?

a. The problem is |X -(-3)| = 6, meaning what numbers are 6 units from -3. Mark the location of -3 **on** the number line below with a "*".

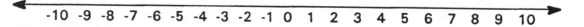

-10 -9 -8 -7 -6 -5 -4 -3 -2 -1 0 1 2 3 4 5 6 7 8 9 10

b. Count off 6 units to the right of -3 and place a ↓ **above** the number line to designate the location of the number that is 6 units away from -3.

c. Count off another 6 units to the left of -3 and place a ↓ **above** the number line to designate the location of the second number that is 6 units away from -3.

d. Record the two values that you determined to be 6 units from -3. ___ and ___

6. Following the steps outlined in #5, use a number line (not the calculator) to solve the following problem: $|X - 2| = 4$
Solution:

$$\longleftarrow \overline{\quad -10 \quad -9 \quad -8 \quad -7 \quad -6 \quad -5 \quad -4 \quad -3 \quad -2 \quad -1 \quad 0 \quad 1 \quad 2 \quad 3 \quad 4 \quad 5 \quad 6 \quad 7 \quad 8 \quad 9 \quad 10 \quad} \longrightarrow$$

The two numbers that are 4 units from 2 are _____ and _____.

7. Repeat this problem, $|X - 2| = 4$, using the INTERSECT option on the graphing calculator to solve for X. Record your screen display for one of the solutions on the graph at the right. Repeat the INTERSECT option and record **both** of the solutions in the blanks below:

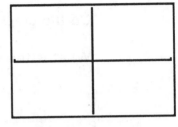

X = ____ or X = ____

8. Do the answers you computed in #7 agree with the answers you determined in #6? If they do not, rework both #6 and #7 to determine the source of your error.

EXERCISE SET

Directions: Graphically solve each of the equations below. Sketch your screen. Circle the points of intersection. Use the INTERSECT option to find the intersections. REMEMBER: Because there are two points of intersection, you will need to repeat the process. Record both of the solutions on the blanks provided.

A. $|2X - 1| = 5$

X = ____ or X = ____

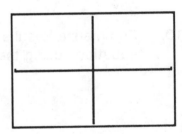

B. $\left|\frac{1}{2}X + 1\right| = 3$

X = ____ or X = ____

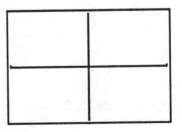

121

C. $\left|\dfrac{4 - X}{2}\right| = \dfrac{8}{5}$

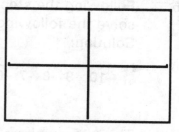

X = ____ or X = ____

D. $|4X + 5| = -2.$

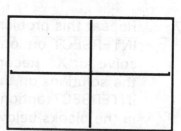

Do the graphs intersect?_____

What is the solution? _____

9. You should be able to determine the solution to $|4X + 5| = -2$ by merely <u>looking</u> at the problem. What clue lets you know that there is no solution? (HINT: Re-read the introductory paragraph to the Unit for the answer.)

NOTE: The equations in this unit were specifically written to conform to the WINDOW values displayed on the screen at the right. If you plan to use your calculator to solve equations in your textbook, you may not be able to see all the intersection points displayed on the screen. You can remedy this problem by adjusting the WINDOW values. Begin by increasing the Xmax and Ymax by 5 and decreasing the Xmin and Ymin by 5 units. (i.e. [-15,15] by [-15,15]) Continue to increase/decrease by increments of 5 until the points of intersection are displayed.

```
███████ FORMAT
 Xmin=-10
 Xmax=10
 Xscl=1
 Ymin=-10
 Ymax=10
 Yscl=1
```

10. Summarize Results: Briefly describe the process of solving absolute value equations using the INTERSECT option.

<u>Solutions:</u> 3e. 6, 3f. 6, 4. X=3 or X=-9, 5d. -9 and 3, 6. -2 and 6, 7. X=-2 or X=6
Exercise Set: A. X= -2 or X=3, B. X=-8 or X=4, C. X=0.8 or X=7.2, D. No, null set

See Unit #1:TI-85 Appendix for discussion of the absolute value function and Unit #10:TI-85 Appendix for discussion of the ISECT (intersection) feature. When there is more than one intersection displayed, you must use the **ISECT** function for each of them. The calculator finds the intersection closest to where the cursor lies before pressing the final **[ENTER]**.

3. Recall that HP 48G by default is set to put each tick mark 10 pixels apart. Be sure that you have changed your PLOT OPTIONS. (See Unit#10 Appendix.)

4. To use ISECT to find each intersection, move the cursor closer to the intersection you want to find and press the ISECT menu key. Pressing [NXT] will redisplay the menu at the bottom of the screen.

UNIT #14
GRAPHICAL SOLUTIONS TO ABSOLUTE VALUE
INEQUALITIES

In the previous unit we looked at solutions to absolute value equations. The equation $|X - (-3)| = 6$ was translated to: "the distance between a number and -3 is 6 units" or "what numbers are 6 units from -3". In this unit we will examine absolute value inequalities and how they relate to distance.

1.   Consider $|X - (-3)| \leq 6$. This inequality will be translated to: find all the numbers whose distance from -3 is 6 units **or** less. We already know the answer to the first portion of the question: find all numbers whose distance from -3 is equal to 6 units. [REVIEW the INTERSECT process from Unit #13 if needed.] You should have found the points of intersection to be $X = 3$ and $X = -9$. These are the solutions to the absolute value **equation** $|X - (-3)| = 6$. Label Y1 and Y2 on the graph at the right.

2.   Now examine the second part of the question: "find all numbers whose distance from -3 is less than 6 units".
     Look at the graph displayed in #1. We know that $Y1 < Y2$ when the graph of Y1 is **below** the graph of Y2. To answer the second part of the question we want to find the X values for which this is true. Press **[TRACE]** and be sure your TRACE cursor is on the graph of Y1.

     As you TRACE along the graph of Y1 you will discover that the portion of Y1 that is less than Y2 is the portion of the graph of Y1 that is below the graph of Y2. This portion has been highlighted on the graph at the right.

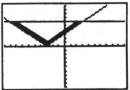

3.   Place your TRACE cursor on the left hand point of intersection and TRACE right along the highlighted portion of the graph of Y1. Observe the X values as you TRACE. Describe what happens to the X values along the highlighted portion of the graph.

4.   On the graph, draw a dotted vertical line from each point of intersection perpendicular to the horizontal axis. Count the "tic" marks on the horizontal axis and label the points where your perpendicular lines touch the axis.

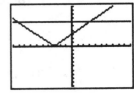

5. The highlighted portion of the graph of Y1 should be between the two points you labeled.

    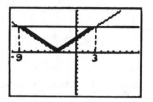

    The solutions to the **inequality** $|X - (-3)| < 6$ are all the X values between -9 and 3 (the critical points).

    The solutions to the **equation** $|X - (-3)| = 6$ are the two points -9 and 3.

    Combining this information, we get the solution to $|X - (-3)| \leq 6$ to be $-9 \leq x \leq 3$.

    Solution Set: $\{X| -9 \leq X \leq 3\}$, Number line graph:

    Interval Notation: [-9,3]

6. Graphically solve $|X - (-3)| > 6$ by following the indicated steps.

    a. Press **[Y =]** and enter abs(X + 3) after "Y1 =" and 6 after "Y2 =".

    b. Press **[GRAPH]**. Sketch the display screen and draw in dotted lines from the intersection points perpendicularly to the horizontal axis.

    c. Use the calculator's INTERSECT option to find the points of intersection.
        X = _____ and X = _____

    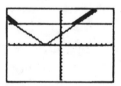

    d. Count the "tic" marks and label the points on the horizontal axis where your perpendicular lines touch the axis.

[NOTE: At this point, the steps listed above are the same steps you used to solve $|X -(-3)| \leq 6$ in #1-5 in this unit and to solve $|X - (-3)| = 6$ in Unit #13.]

    e. The solution is found to be the X values where Y1 > Y2. When the graph of Y1 is above the graph of Y2, as indicated by the highlighted portions, then Y1 > Y2.

    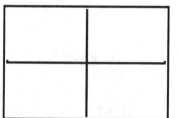

    f. The solutions to the inequality $|X - (-3)| > 6$ are all X values greater than 3 or less than -9: X > 3 or X < -9.

        Solution Set: $\{X \mid X < -9 \text{ or } X > 3\}$

        Number line graph:

        Interval Notation: $(-\infty,-9) \cup (3,\infty)$

[NOTE: -9 and 3 were not included because the inequality states that $|x -(-3)|$ is <u>strictly</u> greater than 6.]

Directions: Graphically solve each of the following inequalities following the steps outlined in #6 above. Record your solution in set notation, number line graph and interval notation.

A. $|2X - 1| \geq 5$

Solution Set:_____

Number line graph: ⟵——————————⟶

Interval Notation:_____

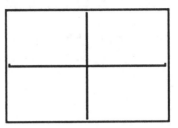

B. $\left|\frac{1}{2}X - 1\right| < 4$

(Be Careful! Enclose the $\frac{1}{2}$ in parentheses.)

Solution Set:_____

Number line graph: ⟵——————————⟶

Interval Notation:_____

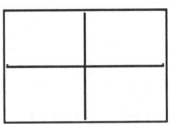

C. $\left|\frac{2X + 5}{3}\right| < 4$

Solution Set:_____

Number line graph: ⟵——————————⟶

Interval Notation:_____

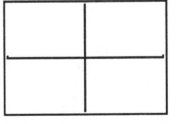

In your own words, explain what type of error could easily be made when graphing the expression $\left|\frac{2X + 5}{3}\right| < 4$ or $\left|\frac{1}{2}X - 1\right| < 4$.

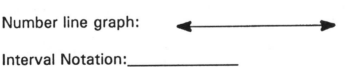

D. $|4x + 2| > -3$
     This particular inequality represents a "special case". Carefully re-TRACE your highlighted portion of the graph before deciding on your solution.

     Solution Set:_____

     Number line graph:

     Interval Notation:_____

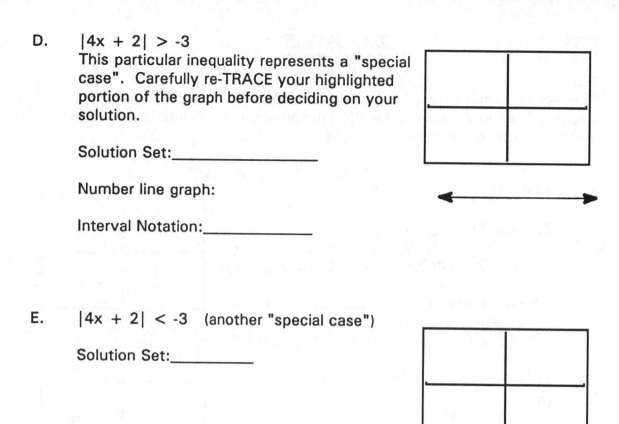

E. $|4x + 2| < -3$ (another "special case")

     Solution Set:_____

F.   Consider why D and E are labelled as "special cases". Could D and E have been solved by merely "looking" at the inequality? Think carefully about the definition of absolute value before formulating your response.

NOTE: The inequalities in this unit were specifically written to conform to the WINDOW values displayed on the screen at the right. If you plan to use your calculator to solve inequalities in your textbook, you may not be able to <u>see</u> the intersection point displayed on the screen. You can remedy this problem by adjusting the WINDOW values. Begin by increasing the Xmax and Ymax by 5 and decreasing the Xmin and Ymin by 5 units, i.e. [-15,15] by [-15,15]. Continue to increase/decrease by increments of 5 until the points of intersection are displayed.

```
WINDOW FORMAT
Xmin=-10
Xmax=10
Xscl=1
Ymin=-10
Ymax=10
Yscl=1
```

You may use the calculator's TEST menu to display a graph that resembles the number line graph of any of the inequalities you have solved. Refer to Unit #12 if you need a review of using the TEST menu.

EXERCISE SET CONTINUED

Directions: Use the TEST menu to display a graph that resembles the number line graph of the solution to each of the inequalities solved in this unit.
i. Sketch the display screen for each problem.
ii. The step number or exercise letter (where the inequality was originally solved) is displayed in parentheses to the right of the inequality. Return to the referenced problem to determine the critical points.
iii. Label these critical points on the display screen sketch.
iv. If the graph displayed does not agree with the solution you determined, then go back and recheck your work.

G. $|x - (-3)| \leq 6$ (#1)   H. $|x - (-3)| > 6$ (#6)

I. $|2x - 1| \geq 5$ (Exercise A)   J. $\left|\dfrac{2x + 5}{3}\right| < 4$ (Exercise C)

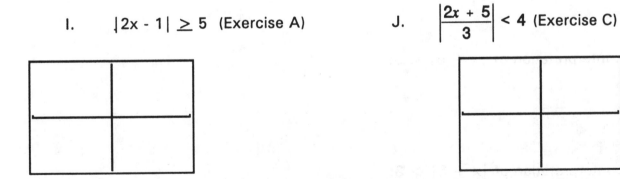

K. |4x + 2| > -3 (Exercise D)   L. |4x + 2| < -3 (ExerciseE)

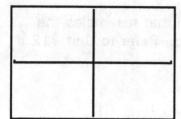   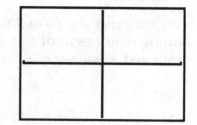

7. In your own words, explain **WHY** the graphical representations of the
    solutions to each of the following problems all look alike.

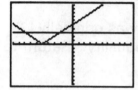   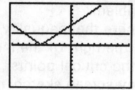   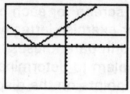

|x + 5| = 3 |x + 5| < 3 |x + 5| > 3

8. Summarize Results: The interpretations of the solutions represented by the
    graphs in #7 are all different. To summarize your results from this unit,
    explain how you interpret the solution represented by each graph.

a. Interpretation of |x + 5| = 3:

b. Interpretation of |x + 5| < 3:

c. Interpretation of |x + 5| > 3:

Solutions: 5. **6c.** X = -9 and X = 3, **6f.**

Exercise Set: A. {X | X ≤ -2 or X ≥ 3}, , (-∞,-2] ∪ [3,∞), **B.** {X | -6 < X < 10}, ,

(-6,10), **C.** {X | -8.5 < X < 3.5}, , (-8.5,3.5), **D.** ℝ or {X | X is a real number}, (-∞,∞)
E. null set

G.   **H.**   **I.**   **J.**   **K.**   **L.**

7. The graph displays all look the same because we are always entering the left side of the equation/inequality at Y1 and the right side at Y2. It is the interpretation of these graphs that yields the correct solution.

Polynomial equations can be solved by the INTERSECT method that was used in previous units. Both sides of the equation can be graphed and the solution(s) determined from the point(s) of intersection. There is, however, another graphic approach to solving equations that will be considered in this unit. This method is the ROOTS method and can be used to find the REAL roots of all the equations you have learned to solve graphically thus far and for any equation you will encounter in the future.

INTERSECT METHOD

First we will review the INTERSECT method. This method is best applied to problems in which the equation is not set equal to zero. Remember, it is critical when using this method that you be able to see the point(s) of intersection that will produce your solutions (or roots).

1.   To solve $2(X^2 - 3) - 2X = 3(X + 2)$ you will need to press **[Y=]** and enter $2(X^2 - 3) - 2X$ after $Y1=$ and $3(X + 2)$ after $Y2=$. Press **[ZOOM] [6:ZStandard]** to display the graphs of the two expressions. One point of intersection is obvious, but the other point is out of the viewing window. (See the screen at the right.) Press **[WINDOW]** and adjust the size of the viewing window. Since the graph appears to have another intersection point "above" the viewing window, we will cursor down with the arrow key and change the Ymax to 20 (an educated guess) instead of 10. Press **[GRAPH]** and your screen should look like the one displayed at the right.

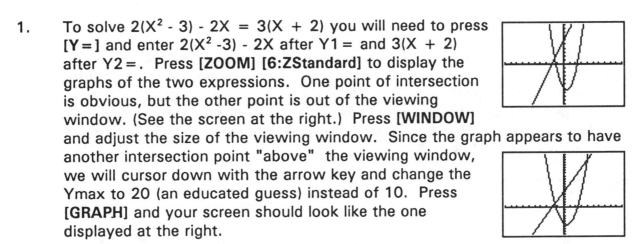

2.   Because the INTERSECT option works independently of the viewing WINDOW, you may adjust your WINDOW as necessary for the given equation. As long as all intersection points are displayed, the calculator will compute the same answer for every calculator, regardless of the WINDOW values.

Use the INTERSECT option (refer to UNIT #10 for a detailed explanation of this process if necessary) twice to compute both roots. Circle the points of intersection. The X values represent the solutions to the equation. Record the solutions: X = _____ and _____.

3. Use the INTERSECT option to solve 2(X + 2)(X - 2) + 4
    = (X + 4)(X - 1) - 2X graphically. After you have entered
    the two expressions after Y1 = and Y2 =, press [ZOOM]
    [6:Zstandard]. Your screen should look like the screen at
    the right. It is difficult to determine the number of points
    of intersection.

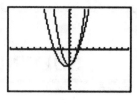

4. To help us determine the points of intersection, we will
    use the ZOOM IN option to get a closer look at the section
    where the two graphs appear to intersect. Press [ZOOM]
    [2:Zoom In]. Use the down arrow key to cursor down to
    the third "tic" mark on the vertical axis (where Y = -3)
    and then use the right arrow key to locate the cursor over
    the graph. (NOTE: The cursor is moved to a point "in the neighborhood" of
    the intersection of the two graphs so the calculator will ZOOM IN on that
    specific region.) Now press [ENTER]. Your graph display should look very
    much like the one at the above right. Two points of intersection are now
    visible.

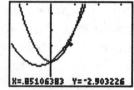

5. Use the INTERSECT option to compute both points of intersection and thus
    determine the two solution values for X (i.e. roots).

    The roots are X = _____ and _____.

**TROUBLE SHOOTING NOTE: When using the ROOT option, the calculator
determines the root(s) by locating the X-intercept(s). At the X-intercept(s) the graph
is crossing the horizontal axis (X-axis) and the Y value will be ZERO. There will be
times when the calculator will be very close to ZERO but will not display ZERO
exactly. For example Y = 6 E -9 would be Y = .000000006, which is for all
practical purposes a ZERO.**

ROOTS METHOD

There is an alternate method for solving equations instead of the INTERSECT option.
This method is called the zeroes or ROOTS option. If an equation is set equal to
zero then when you graph the non-zero side of the equation the real roots are at the
point(s) where the graph crosses the horizontal axis (X-axis). This method is perfect
for equations (of any type) that are already set equal to zero. (In fact, it is the
preferred method for equations in that form.)

6. If an equation is not set equal to zero, then you should try the INTERSECT
    method first. WHY? Because often when you begin to "work" on an
    equation with paper and pencil you make careless errors. By merely entering
    the existing equation into the calculator you run less risk of error. If the
    points of intersection are not clearly visible then you can set the equation
    equal to zero and use the ROOTS option.

7. Solve the same equation that was in #3, by the ROOTS method. Begin by setting $2(x + 2)(x - 2) + 4 = (x + 4)(x - 1) - 2x$ equal to zero. We will perform the **least** amount of algebra possible to accomplish this:

$2(x+2)(x-2)+4 = (x+4)(x-1)-2x$

$2(x+2)(x-2)+4 \underline{+ 2x} = (x+4)(x-1) - 2x \underline{+ 2x}$ a. add 2x to both sides

$2(x+2)(x-2)+4 + 2x = (x+4)(x-1)$

$2(x+2)(x-2)+4 + 2x \underline{- (x+4)(x-1)} = (x+4)(x-1) \underline{- (x+4)(x-1)}$ b. subtract (x+4)(x-1) from both sides

$2(x+2)(x-2)+4 \underline{+ 2x - (x+4)(x-1)} = 0$ c. compare the underlined part to the original equation

It is this algebraic process of setting the equation to zero that encourages us to use the INTERSECT option if at all possible when the equation is not already set equal to 0!

8. Enter the NON-zero side of the equation after Y1 =. Press **[ZOOM] [6:Zstandard]** to automatically set the standard viewing WINDOW. The graph is displayed at the right.

The curve appears to "dip" slightly below the horizontal axis. To get a better view of this section of the curve, we will need to alter the WINDOW values.

Press **[WINDOW]**, use the down arrow key to move down to Xmin and change the value to -3, then move down to Xmax to change the value to 3. The numbers -3 and 3 were selected by counting the "tic" marks on the horizontal axis to determine a satisfactory distance <u>away</u> from the section that "dips" below the axis. Set the Ymin at -3 and the Ymax at 3. This centers the axes. Press **[GRAPH]**.

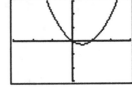

NOTE: You could also have used the ZOOM IN option (just like the ZOOM OUT) here instead of changing WINDOW values.

9. To use the ROOTS option, press **[2nd] <CALC> [2:root]**.

a. **Set lower bound:** The screen display asks for a lower bound. A lower bound is an X value smaller than the expected root; move your cursor to the left of the left-hand root and press **[ENTER]**. Because our roots are determined on the horizontal axis, a lower bound is always determined by moving your cursor to the <u>left of the root</u>. Notice at the top of the screen a ▸ marker has been placed to designate the location of the lower bound.

b. Set upper bound: Similarly the upper bound is always determined by moving the cursor to the <u>right of the root</u>. At the upper bound prompt, move the cursor to an X value larger than the expected root (<u>**DO NOT**</u> go past the right-hand root) and press **[ENTER]**. Again, a ◄ marker is at the top of the screen to designate the location of the bound.

NOTE: If your bound markers do not point toward each other, ► ◄, then you will get an "ERROR:bounds" message. If this happens, you will need to start the ROOT calculation over.

c. Locate first root: Move the cursor to the approximate location where the graph crosses the X-axis for your guess. When you press **[ENTER]** the calculator will search for the root, within the area marked by ► and ◄. The root is X = 0.

NOTE: Your calculator should display X = 0   Y = 0, which means that the expression entered at **Y=** has a value of 0 when X = 0. If your display is X = 1.01 E-14 and Y = 0, your calculator has determined the root X to be a value very close to zero. To verify that Y = 0 when X = 0, scroll through the TABLE to X = 0 and see that Y = 0.

d. Locate subsequent roots: Repeat the entire process outlined above to determine the right-hand root. This root is X = 1.

EXERCISE SET

DIRECTIONS: Solve each of the following quadratics, using the ROOT option. So that everyone's graph will look alike, press **[ZOOM] [6:Zstandard]**. Sketch your graph display, circle the two roots and record their values in the blanks provided. Beneath each problem, factor the quadratic that you graphed.

A. $2X^2 + 7X - 15 = 0.$

Factorization_____

The roots are X = _____ and X = _____

B. $X^2 + 8X - 9 = 0$

Factorization:_____

The roots are X = _____ and X = _____.

C. $X^2 - 25 = 0$

Factorization:_____

The roots are X = _____ and X = _____.

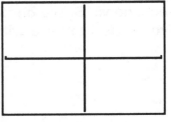

D. $(X-2)^2 + 3X - 10 = 0$

Factorization:_____

The roots are X = _____ and X = _____.

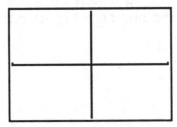

E. $X^2 + 6X + 9 = 0$

Factorization:_____

The root(s) is/are X = _____ .

(Note: the point at which the graph touches, but does not cross the X-axis, produces two identical roots - often called a double root.)

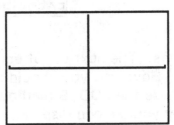

F.   Compare the factorization of your polynomial to the real roots that you determined in each of the problems above. How do they compare?

G.   If you know that the roots to an equation are 4 and -2, you should be able to write an <u>equation</u>, in factored form, based on the information gathered in F. Write an <u>equation</u>, in factored form:

_____

10. The equations that you have solved thus far in this unit have all been second degree equations. Based on the exercises you have worked, how many roots should you expect to have with a second degree quadratic equation?

11. **ONE** of the equations does not conform to the pattern. Which one is it and why is the number of roots different from the rest of the problems?

EXERCISE SET CONTINUED

Directions: The next set of equations contain polynomials that are not second degree. However, you should be able to factor these polynomials.
i. First use the ROOTS method to solve the equation.
ii. Copy your screen display.
iii. Circle the real roots.
iv. Record the value of the roots in fractional form.
v. Factor the polynomial.

H. $X^3 - 7X^2 + 10X = 0$

Factorization:_____

The roots are X = _____, _____ and _____.

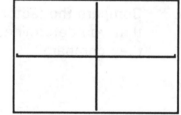

I. $X^4 - 5X^2 + 4 = 0$

Factorization:_____

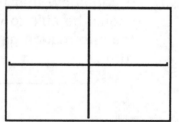

The roots are X = _____,_____,_____ and _____.

138

J. $3X^4 + 2X^3 - 5X^2 = 0$

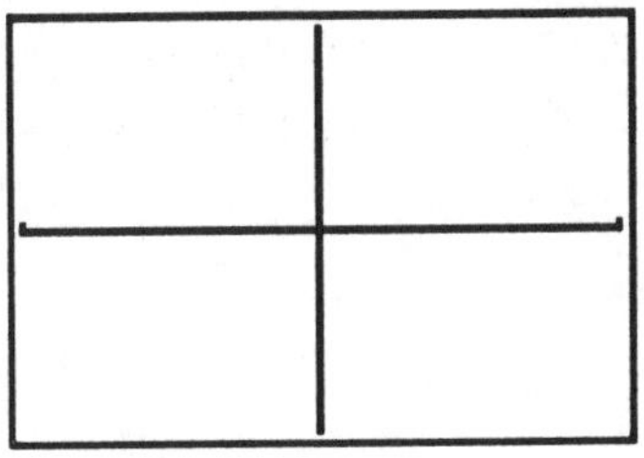

Suggestion: Change the WINDOW values as in #8, only this time use -5 and +5.

Note: There is a double root in this problem just like exercise E. Which root is the double root?_____

Factorization:_____

The roots are X = _____,_____ and _____.
Which root is the double root? _____

12. What conclusions can you draw about the number of real solutions and the degree of the polynomial equations you have been solving?

13. Summarize Results: When summarizing what you have learned in this unit, you should address the following:
a. how to use the ROOT method,
b. the relationship between a polynomial's roots and its factors,
c. the relationship between the degree of a polynomial and the number of roots it has and
d. how many roots occur when the graph is tangent (touches but does not cross) the horizontal axis.

14. Highlight #4 and #9 of this unit for your QUICK REFERENCE list of operations.

Solutions: **2.** -1.5 and 4, **5.** 0 and 1,
Exercise Set: A. $(2X-3)(X+5)=0$, $X=-5$ and $X=1.5$,
B. $(X+9)(X-1)=0$, $X=-9$ and $X=1$ **C.** $(X-5)(X+5)=0$, $X=-5$ and $X=5$, **D.** $(X-3)(X+2)=0$, $X=-2$ and $X=3$, **E.** $(X+3)(X+3)=0$, $X=-3$ **G.** $(X-4)(X+2)=0$
10. two **11.** Letter E. We did not see two distinct real roots, but rather two identical roots (often called a double root).
H. $X(X-5)(X-2)=0$, $X=0,2$ and 5 **I.** $(X-2)(X+2)(X-1)(X+1)=0$, $X=2,-2,1$ and -1
J. $X^2(3X+5)(X-1)=0$, $X=0,-5/3$ and 1, Zero is the double root.
12. The number of solutions is the same as the degree of the equation. However, not all the solutions are distinct (different). Some appear as multiple roots.

140

ROOTS METHOD

The TI-85 has a ROOTS option, however it is accessed in a different manner than the TI-82. First graph the desired function. Then with the graph and standard graphing menu displayed, select [MORE] and choose [F1](MATH). Then you simply select [F3](ROOT), place the cursor on the graph near the desired root, and press [ENTER]. Your calculator will display the word "ROOT" and the x and y coordinates of the root. (Of course the y-coordinate *should* be zero!) Should you desire to restrict the region you are examining for roots, you may set the LOWER and UPPER bounds under the MATH menu, but this is seldom necessary.

The POLY Function

The TI-85 is equipped with an additional tool for finding the roots of a polynomial, accessed by selecting [2nd] <POLY> (located above the [PRGM] key). To use this tool, you must have the desired polynomial in the form

$$a_n x^n + a_{n-1} x^{n-1} + a_{n-2} x^{n-2} + \ldots + a_2 x^2 + a_1 x + a_0 = 0$$

In other words, you must multiply out all the factors and collect like terms, then arrange the polynomial in decreasing order of powers. After doing so, the highest exponent in the polynomial (the first listed if the terms are in decreasing order of powers) is the order of the polynomial. Then simply enter the coefficients and choose [F5](SOLVE) to see the solutions.

Example: Find all solutions to $2x(x^2 - 7) = 3(x^2 - 5)$.

Solution: First multiply out all the terms and collect them on the left-hand side, setting the polynomial equal to zero and writing it in decreasing order of powers to get:
$$2x^3 - 3x^2 - 14x + 15 = 0$$

Now press [2nd] <POLY>. This is a third order, or third degree, polynomial. Type a [3] and press [ENTER].

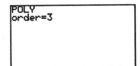

Enter the coefficients of the polynomial, being careful with plus (+) and minus (-) signs:

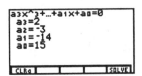

141

Finally, push [F5](SOLVE) to see the roots:

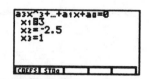

CAUTION: One thing to be aware of is that many polynomials have no real roots, or have some real and some complex roots. The TI-85 simultaneously finds all complex and real roots and displays them. For example, consider $x^3 - x^2 + 4x - 4 = 0$. This has only one real root, namely $x = 1$, yet the TI-85 gives the following:

The TI-85 displays complex numbers **a + bi** in the form **(a,b)**. Thus a root of (-4,0) would be -4 + 0i or -4, a real root. If you are asked to find *real* roots, you must select *only* those roots whose second coordinate (the complex part) is zero. In this case, you will look at the pair (1, 0), and see that the first coefficient, the "1", is your only real root, since it is the only pair with a "0" as the second coefficient. The other two pairs are the complex numbers 2i and -2i. (The first coordinate, -1E-13, is essentially zero.) Consult the TI-85 GUIDEBOOK or your instructor for further instruction on using the POLY function.

UNIT #15 APPENDIX
Casio fx7700GE

The Casio does not have a **ROOTS** feature. However, it does have the quadratic formula built into it. The quadratic formula is used to solve quadratic equations in the form $ax^2 + bx + c = 0$. To access this feature, select **[EQUA]** from the MAIN MENU, then select **[F3]** for **[PLY]**. Enter the coefficients a, b, and c, and press **[F1]** for **[SOL]**.

UNIT #15 APPENDIX
HP 48G

ISECT METHOD

Use the ISECT command to find each intersection. To "zoom in", select ZOOM menu key while in the PICTURE environment then press (ZIN).

ROOT METHOD

The ROOT option on the HP 48G is accessed by pressing (FCN) to open the FUNCTION menu while in the PICTURE environment. Press the menu key corresponding to ROOT to find the value of the root closest to the current cursor position. You do not set upper or lower bounds like in the TI-82.

In the previous unit we solved factorable polynomial equations. Several observations can be made from that unit: 1) for each factor there was a root 2) the number of factors corresponded to the degree of the polynomial 3) the number of real roots was equal to or less than the degree of the polynomial. CONCLUSION: A polynomial equation of degree n will have <u>at most n real roots</u>. This unit will examine polynomial equations, specifically focusing on quadratics, that do not factor and hence may not have any real roots, or at best roots that are irrational.

1.   Consider the equation $3x^2 - 18x + 25 = 0$. This trinomial does not factor, so you will have to solve using either the "completing the square" method or the quadratic formula.

a. <u>QUADRATIC FORMULA:</u>

$$3x^2 - 18x + 25 = 0$$

$$a = 3,\ b = -18,\ c = 25$$

$$\frac{-b \pm \sqrt{b^2 - 4ac}}{2a} \approx 3.82 \ \text{ or } \ 2.18$$

(Use the STO▶ feature on your calculator to evaluate.)

b. <u>COMPLETING THE SQUARE:</u>

$$3x^2 - 18x + 25 = 0$$

$$\frac{1}{3}(3x^2 - 18x + 25) = \frac{1}{3}(0)$$

$$x^2 - 6x + \frac{25}{3} = 0$$

$$x^2 - 6x = -\frac{25}{3}$$

$$x^2 - 6x + 9 = -\frac{25}{3} + 9$$

$$(x - 3)^2 = \frac{2}{3}$$

$$x - 3 = \pm\sqrt{\frac{2}{3}}$$

$$x = 3 \pm \sqrt{\frac{2}{3}}$$

$$x = 3 \pm \frac{\sqrt{6}}{3}$$

145

2. Enter each of the two roots found in #1b into the calculator to determine a decimal approximation. Record the two roots **exactly** as they are displayed on your screen. DO NOT round off.

X = _____ and X = _____

3. Enter $3X^2 - 18X + 25$ after Y1 = and graph in Zstandard (press **[ZOOM] [6:Zstandard]**). Use the ROOT option twice to compute both roots of the equation. The screens displayed indicate both roots. These roots should be comparable to the values you determined in #2.

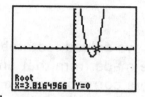 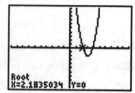

EXERCISE SET

Directions: Use the ROOT option to find the **REAL** roots of the following quadratic equations. Sketch the screen display and record your solutions.

A. $-5X^2 + 5X + 8 = 0$

X = _____ and _____

What happens when you try to convert your roots to a fraction?

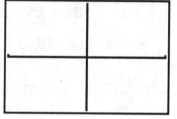

Why?

Solve by either completing the square or using the Quadratic Formula. Approximate your solutions and compare them to the calculator answers you have recorded above. They should be the same.

B. $X^2 - 6X + 4 = 0$

X = _____ and _____

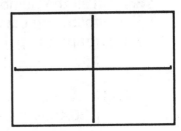

C. $3X^2 + 18X - 3 = 0$

X = _____ and _____

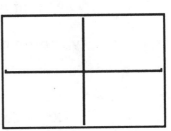

D. $X^2 + 5X + 8 = 0$

Can we find the roots of this equation using the TI-82?_____

Why not?

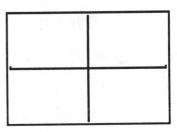

Solve "by hand" using either the Quadratic Formula or by completing the square.

X = _____ and _____

Graphs which intersect the X-axis will have real roots because the X-axis represents the set of real numbers. Roots which are complex numbers will not be represented on the X-axis. Thus an equation with complex roots will not intersect the X-axis. There will, however, be two roots because complex roots always occur in conjugate pairs.

E. Tell the number and type of roots (real or complex) each of the following equations will have. DO NOT SOLVE the equations, simply graph the polynomial function and check the number of X-intercepts - if any.

i. $6X^2 + 2X - 4 = 0$

Number of roots:_____

Type of roots:_____

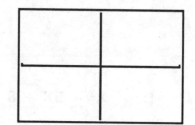

ii. $2X^2 + 5X + 5 = 0$

Number of roots:_____

Type of roots:_____

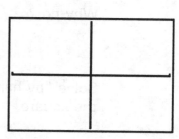

iii. $X^2 + 6X + 9 = 0$

Number of roots:_____

Type of roots:_____

4. Summarize Results: On the next page, write a summary of what you learned in this unit. You should address the following:
a. How to use the calculator to find real roots of quadratic equations.
b. The relationship between the graph of the quadratic and the number and type of roots.

Solutions: **2.** X = 3.816496581 and X = 2.183503419 Exercises: **A.** X = -.860147 and
X = 1.8601471, These answers will not convert to fractions because they are irrational numbers.
B. X = .76393202 and X = 5.236068 **C.** X = -6.162278 and X = .16227766 **D.** No, It does not

intersect the X-axis and therefore has no real roots. $X = \dfrac{-5 \pm i\sqrt{7}}{2}$ **E. i.** 2, real **ii.** 2, complex

iii. 1, real

149

150

1. A travelling circus has a "human cannonball" act as its grand finale. The equation $Y = -.01X^2 + .64X + 9.76$, where Y = height in feet and X = horizontal distance travelled in feet, represents the flight path of the human cannonball. Display a graphical representation of this equation and use the TRACE feature to answer the questions.

    a. TRACE along the graph in ZStandard (press [ZOOM] [6:ZStandard] to automatically set this WINDOW) and examine the numbers displayed at the bottom of the screen.
    What do the X values represent?

    What do the Y values represent?

    b. In order to have "friendly" values for the X and Y, we will change the viewing WINDOW to ZInteger. Press [ZOOM], cursor down to highlight 8 for ZInteger and press [ENTER]. Be sure your cursor is at X = 0 and Y = 0 (use the arrow keys to move the cursor to the point where the axes intersect to display X = 0 and Y = 0) and press [ENTER] again to set the viewing WINDOW to ZInteger. We moved the blinking cursor to X = 0 and Y = 0 to keep the axes centered on the screen when we changed to ZInteger. When changing to ZInteger, you may place the cursor at any position on the screen and the axes will intersect at that point. Now TRACE along the curve and observe the "friendly" values represented at the bottom of the screen.

    NOTE: ZInteger yields integer values for X when tracing on the graph. This is particularly valuable when X only has meaning as an integer value.

    c. This curve represents the flight path of the human cannonball. **TRACE** along the path to the right. How far, <u>approximately</u>, has the human cannonball travelled horizontally when he hits the ground?_____

    d. The human cannonball is travelling at speeds up to **65 mph**. To land on the ground would mean certain death. If he uses a net for his landing, how far will he have travelled horizontally if the net is 11 feet above the ground?_____

    e. What is the maximum height that he reaches during the course of his flight?_____

f. How far has he travelled horizontally when he reaches this maximum height?_____

NOTE: The actual distance he has travelled is a length of arc along the curve of the parabola. To calculate this length requires the use of calculus.

g. The human cannonball is shot out of the cannon head first, so all of the distances are measured from his head. TRACE along the curve to X = 0 and Y = 9.76. Explain the meaning of these two values.

2. Blaire is a pitcher for the Girls Slowpitch softball team at her middle school. The height of the softball X seconds after she releases a pitch is given by the formula h = -16X^2 + 30X + 3.

    a. Find the length of time it will take the ball to hit the ground if the batter swings and misses.

    <u>Solution:</u> When the ball hits the ground the height will be h = _____. Thus the equation we are trying to solve for X is 0 = -16X^2 + 30X + 3. We want to solve this equation using the roots feature, so enter -16X^2 + 30X + 3 after Y1 = and press [ZOOM] [6] to display the graph in the standard viewing WINDOW. You will need to adjust the viewing rectangle!

    Press [TRACE] to approximate the width of the graph at the points it crosses the X-axis (it appears to cross at about -.1 and 2) and to determine its approximate height (the highest y value is about 16.9). Based on this information, press [WINDOW] and change the Xmin to -.1, Xmax to 2.25, Ymin to -5 and Ymax to 20. Press [TRACE] to simultaneously view the graph and to activate the TRACE feature. TRACE around the curve to see if it is possible to determine the exact time the ball will be 0 feet from the ground. TRACE will not yield the exact answer. We now use the ROOT option to determine the two roots of the equation.

    X = _____ or X = _____

    One of these roots is not valid. Which one is it and why?

    If you are having difficulty answering "why" then ask yourself this question: Can X equal a negative number? Remember, X represents time.

152

After using the ROOT option on the valid root, you should have determined X to be approximately 1.9701695. This means that it will take the ball approximately 1.97 seconds to hit the ground if the batter misses.

b. What is the highest point that the ball reaches during the pitch?

Solution: Again we can TRACE to find the highest point, but letting the calculator determine the MAXIMUM point will be more accurate. Press [2nd] <CALC> [4:maximum]. The calculator is now ready to determine the maximum point on the curve. Set upper and lower bounds as you do when using the ROOT option. Remember, the X value displayed represents seconds elapsed since the pitch was thrown and the Y value indicates the height of the ball at that point in time. Y = _____ indicates that the ball reached a maximum height of approximately _____ feet.

c. How long does it take the ball to reach its maximum height?_____

d. What is the height of the ball 2.1 seconds after the pitcher releases it? (TRACE along the path of the curve to X = 2.1 seconds.) Carefully explain your answer.

e. Use the TRACE cursor to TRACE along the path of the curve from left to right. Explain, in your own words, what information the X and Y values at the bottom of the screen are giving you.

f. Does the curve represent the path of the ball in flight? If you answer yes, then explain which part of the graph display represents the distance the ball travels.

If you answer no, explain why not.

3. Bridges are often supported by arches in the shape of a parabola. The equation $y = \frac{10}{7}x - \frac{2}{49}x^2$, where Y = height and X = distance from the base of the arch, provides a model for a specific parabolic arch that supports a bridge. Will this arch be tall enough for a road crew to build a county road under?

Solution:

Setting the viewing WINDOW

a. Begin by entering the polynomial at the Y1 = prompt and press [ZOOM] [6:ZStandard] to graph in the standard viewing rectangle.

b. This curve represents the support to a bridge. You need to be able to see the entire curve (i.e. support). TRACE along the curve, recording the following (to the nearest integer): left most X-intercept, maximum Y value, and right most X-intercept.

left most X-intercept:_____   maximum Y value:_____

right most X-intercept:_____

c. Use the above information to set our WINDOW values so that the entire graph is displayed. Press [WINDOW] and set the WINDOW values as follows: Xmin = -1, Xmax = 37, Xscl = 1, Ymin = -1, Ymax = 13, Yscl = 1

These values were selected to ensure that we can see slightly beyond the perimeter of the graph displayed.

NOTE: For more information on setting viewing WINDOWS, refer to the unit entitled: "Where Did the Graph Go?".

Solving the problem

d. Begin by determining the height (to the nearest tenth) of the highest point under the arch.

Maximum height = _____

e. If the average vehicle is no more than 6 feet high, can the vehicle drive under the arch?_____ Explain how you determined your answer.

f. If a two lane road is 20 feet wide, will it fit between the bases of the arch?_____ Explain how you determined your answer.

154

g. Can the average vehicle drive in either lane under the arch and not scrape the paint off the roof? (or scrape the roof off the car??) That is to say, if this 20 foot wide road is centered under the arch, is the arch at least 6 feet above the road at all points in its width?

4. The local community theater is considering increasing the price of its tickets to cover increases in costuming and stage effects. They must be careful because a ticket price that is too low will mean that expenses are not covered and yet a ticket price that is too high will discourage people from attending. They estimate the total profit, Y, by the formula $Y = -X^2 + 35X - 150$, where X is the cost of the ticket.

Before attempting to graphically solve the problem, set your viewing WINDOW by following steps a - c in #3 above. **MAKE SURE THAT YOU CAN SEE THE ENTIRE CURVE.**

a. What is the maximum amount that can be charged for a ticket to maximize the profit?_____

b. What is the maximum profit?_____

c. If $13 dollars is charged for each ticket, what will the profit be? (What are your solution options here? We could return to the home screen and evaluate the polynomial for X = 13 by using the **STO**re feature or we could scroll through the TABLE in search of X = 13. However since we have been using the CALC menu to investigate the graph, we will look at **value (EVAL X)** which is the first entry option under the CALC menu. Access the CALC menu and press [1] to select value and display the **EVAL X** prompt. At the prompt, enter 13 and press [ENTER]. What will the profit be when $13 is charged for each ticket? _____

BEWARE: EVAL X only works when the value selected for X is between the Xmax and Xmin values on the WINDOW screen.

d. If they predict a profit of $150.00 on a play, how much was charged per ticket? (Scroll through the Y values in the TABLE to answer this question.)_____

e. When 0 tickets are sold (X = 0) explain the meaning of the Y value displayed on the screen.

f. Use the TABLE display to determine at what point the theater "breaks even", i.e. How many people must attend for there to be no money lost and yet no profit made?

_____

5. Summarize Results: Write a summary of what you learned in this unit. You should address the following:
   a. how and when to adjust WINDOW values,
   b. use of the MAXIMUM option, and
   c. use of the value (EVAL X) option, including its restrictions.

6. Highlight #2b (Maximum) and #3c (EVAL X) of this unit for your QUICK REFERENCE list of operations.

Solutions: 1c. approx. 76.5 ft. **1d.** 62 ft. **1e.** 20 ft. **1f.** 32 ft. **1g.** It means that his head is 9.76 feet above the ground before he is shot from the cannon. **2a.** h = 0
2b. Y = 17.0625, 17 feet **2c.** approx. 1 second **2d.** Y = -4.56, the ball is 4.56 feet into the ground. **2e.** The X values represents the time (in seconds) that the ball is in the air; the Y value represents its height. **2f.** Yes: This is obviously an incorrect answer, because nothing represents distance. No: This is the correct response, because the graph is relating the time (X) to the height of the ball (Y). **3d.** 12.5 **3e.** Yes **3f.** Yes, because the supports are 35 feet apart. **3g.** Yes **4a.** $17.50 **4b.** $156.25 **4c.** $136 **4d.** 15, 10 **4e.** They have lost $150 in preparation cost. **4f.** The theater breaks even when tickets are priced at $5 each or $30 each.

156

UNIT #17 APPENDIX
TI-85

To find maximum values using the TI-85, first graph the function, then with the graphing menu at the bottom press **[MORE]**. Press **[F1](MATH)** then **[MORE]** **[F2](FMAX)**. Place the cursor near the maximum point and press **[ENTER]**. The calculator will display the maximum value.

To access the EVAL feature on the TI-85, graph the function and on the GRAPH menu at the bottom, press **[MORE]** twice and select **[F1](EVAL)**.

The Casio does not have a MAXIMUM or a MINIMUM feature or an EVAL feature.

1.   b. In the PICTURE environment press **(ZOOM)** to bring up the ZOOM menu on the display. Press **[NXT]** twice so that ZINTG appears on the menu. Press the key corresponding to ZINTG. Unlike the TI-82, the graph will not scroll as you use TRACE. Therefore, in order to see the graph, you must either change the horizontal and vertical ranges or "zoom out." You can also just "zoom out" horizontally or vertically using the HZOUT or VZOUT commands in the ZOOM menu. Another option is to move the cursor and select the CNTR command in the ZOOM menu. This will maintain the size of the screen display, but center the display where the cursor was placed.

2.   b. We can let the calculator determine the MAXIMUM point. In the PICTURE environment, press **(FCN)** to display the function menu, then press **(EXTR)** to find the maximum. EXTR is an abbreviation for extremum. By pressing **(EXTR)** the calculator finds the closest maximum or minimum.

4   c. To evaluate the polynomial for X = 13, you must return to the HOME screen and use the VAR menu to define X to be 13 then evaluate the polynomial. Note that if you plotted the polynomial directly instead of defining the variable (see Unit#10 Appendix), EQ is a variable for the polynomial that you just plotted.

Previously, absolute value inequalities such as |X + 3| < 5 were solved by graphing the left side of the inequality at Y1 =, the right side at Y2 = and then determining where Y1 was graphically less than Y2. The observation was made that Y1 < Y2 wherever the graph of Y1 was below (less than) the graph of Y2. That is to say, the X values where Y1 < Y2 are the solutions to the inequality. This unit will examine polynomial inequalities, using what was learned about graphical solutions of absolute value inequalities in conjunction with the ROOT option. REMEMBER: To use the ROOT option, the equation - or inequality- must have a zero on one side of the inequality.

SPECIAL CASES

The first four examples represent "special cases". They will be the quickest to solve of all the inequalities. All graphs will be displayed on the ZStandard WINDOW unless otherwise noted. Set this viewing window now by pressing [ZOOM] [6:ZStandard].

1. To graphically solve $2X^2 - X + 1 > 0$ we let Y1 represent the left side and Y2 the right side. We want to know <u>where</u> Y1 > Y2. Enter $2X^2 - X + 1$ at the Y1 = prompt and 0 at the Y2 = prompt. Press [GRAPH]. Since Y2 = 0 is the X-axis, we do not "see" it as a separate line. It is, however, graphed. (Your display should correspond to the display at the right.)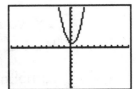

    Press [WINDOW], use the right cursor to highlight FORMAT, cursor down to AxesOn and right to AxesOff, and press [ENTER]. Now press [GRAPH]. It is now obvious that the X-axis and Y2 = 0 are the same line. From this point on, we will make a mental note that Y1 is always being compared to the X-axis and we will not enter Y2 = 0. <u>Think</u> about where Y1 is greater than Y2 (i.e. the X-axis). It is greater than the X-axis where it is **above** the axis. Use your highlighter pen to highlight the portion(s) of Y1 that are above the X-axis. Since all portions of Y1 are greater than the X-axis and since any real number is an acceptable value for X, we can conclude that the solution set is ℝ (the set of real numbers).

    Before proceeding, turn the axes back on (press [WINDOW], access FORMAT, cursor down to AxesOn and press [ENTER]).

2. What is the solution if the inequality is $2X^2 - X + 1 < 0$ instead? Examine the graph displayed in #1. Where is $2X^2 - X + 1 < 0$ (i.e for what values of X is the graph below the X-axis)?

3. Consider the graphical solution to $X^2 + 4X + 4 \leq 0$. The algebraic statement says that the trinomial is less than **OR** equal to zero. Remember, 0 is represented by the X-axis. Enter $X^2 + 4X + 4$ after Y1 and sketch the graph that is displayed. **TRACE** along the path of the curve.

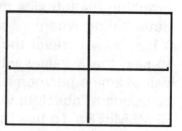

   a. At what point(s) is the graph of $X^2 + 4X + 4$ **LESS THAN** 0?_____

   In other words, when TRACING, for what X-values are the corresponding Y-values negative?

   b. Use your ROOT option to determine where the graph is **EQUAL TO** 0._____

   Although we were solving an inequality, there is only **ONE** valid solution: $X = -2$.

4. a. What would the solution set be if $X^2 + 4X + 4 \geq 0$?_____ Explain why.

   b. What would the solution set be if $X^2 + 4X + 4 > 0$?_____ Explain why.

162

In the above case of $X^2 + 4X + 4 \leq 0$ the critical point in the solution set was $X = -2$. At $X = -2$, $X^2 + 4X + 4 = 0$. Negative two is a root of the equation. When solving inequalities, the critical points (i.e. the roots of the corresponding equation) will be the boundary values of the interval(s) of the solution region(s). The following example illustrates the procedure for solving quadratic inequalities. The way we <u>interpret</u> the graphical display is what determines the solution to a given equation or inequality.

5. We will illustrate the steps for finding the solution set of an inequality by solving $X^2 - X - 6 < 0$:

a. Be sure that the right hand side of the inequality is a ZERO.

b. Enter the polynomial $X^2 - X - 6$ at the Y1 = prompt.

c. On the display at the right sketch the graphical representation of the solution.

d. Circle the X-intercepts (i.e. zeroes or roots) of the graph. These are the critical points. The circles will remain OPEN because the strict inequality symbol, "$<$", indicates that the critical points are not to be included as part of the solution set.

e. Use the ROOT option (under the "Calc" menu) to determine the values of the X-intercepts. Label the values of these two critical points on the display.

f. Use your highlighter pen to highlight the section of the graph of $X^2 - X - 6$ that is **LESS THAN** 0 (i.e. below the X-axis).

g. The solution set is the set of all X values that yield the highlighted section of the graph. That is, $\{X | -2 < X < 3\}$. Represented on a number line graph:

Alternate Option: The TABLE feature of the TI-82 is helpful in determining solutions, once the critical points have been calculated.

Set your table to begin at a minimum of -2 (-2 was chosen because it is the critical point furthest to the left) with increments of 1. Press [2nd] <TblSet> to access this menu.

Access the table by pressing [2nd] <TABLE>. Use the ▲ arrow key to scroll through values of X that are less than -2. As you are scrolling, you should observe the corresponding Y1 values. Notice that they are positive.

Because we are concerned with solutions to the inequality $X^2 - X - 6 < 0$ and have entered $X^2 - X - 6$ at the Y1 = prompt, we now know that when $X < -2$, $Y1 > 0$, Therefore, these values to the left of our critical point are NOT part of our solution.

Use the ▼ arrow key to scroll through values of X that are between our critical points of -2 and 3. Notice that the Y1 values between -2 and 3 are negative, and therefore ARE solutions to our original inequality. We can therefore shade between -2 and 3 on the number line graph. Notice that the Y1 values equal 0 at the critical points, and are NOT solutions to our inequality.

As we scroll through values greater than 3, we notice that the corresponding Y1 values are positive, and are therefore NOT part of our solution set.

6.   Use the steps outlined in #5 (or the TABLE) to help you solve $X^2 - X - 6 \geq 0$.
Sketch the graph that you display, label the critical points, highlight the solution region(s), and interpret the solution.

Record your solution set:_____

and display this information as a number line graph:

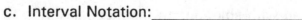

7.   How was the problem in #6 different from #5?

EXERCISE SET

Use the steps outlined in #5 to graphically solve each of the following inequalities. For each problem you **must** sketch the graphical display and label the critical points. Record your solution in the following forms:
a. solution set   b. number line graph   c. interval notation

A. $2X^2 - X - 10 \leq 0$

      a. Solution Set:_____

      b. Number line graph: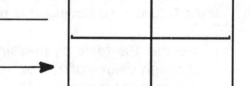

      c. Interval Notation:_____

164

B. $3X^2 + X - 4 \geq 0$ (Record critical points as fractions.)

   a. Solution Set:_____

   b. Number line graph:

   c. Interval Notation:_____

C. $3X^2 + X - 4 < 0$ (Record critical points as fractions)

   a. Solution Set:_____

   b. Number line graph:

   c. Interval Notation:_____

D. $X^2 + 10X + 25 \geq 0$

   a. Solution Set:_____

   b. Number line graph: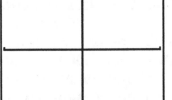

   c. Interval Notation:_____

E. $X^2 + 10X + 25 > 0$

  a. Solution Set:_____

  b. Number line graph: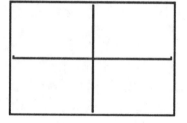

  c. Interval Notation:_____

F. $4X^2 - 12X < -9$

  a. Solution Set:_____

G. $4X^2 - 12X + 9 > 0$

  a. Solution Set:_____

  b. Number line graph:

  c. Interval Notation:_____

(QUESTION: Is the critical point actually 1.4999998 or 1.5? Attempting to convert to a fraction would seem to indicate that it is not 1.5. However, since the critical point(s) is the solution to the equation $4X^2-12X+9=0$, then the critical point should be the X value in the TABLE where Y1 = Y2. This can be accomplished in two ways:

1. Increment the TABLE by 0.5, instead of 1, since we want to know if X = 1.5. Press [2nd] <TblSet>, make this adjustment and then press [2nd] <TABLE>. Scrolling to X = 1.5 we see that Y1 = Y2. Thus X = 1.5 is the exact critical point; X = 1.4999998 is an approximation.

2. Use the **Ask** feature of the TABLE. Press [2nd] <TblSet>, cursor down to **Indpnt** and over to **Ask** and press [ENTER]. Press [2nd] <TABLE>, enter 1.5 at the X= prompt and press [ENTER].

H. Summarize Your Results: Your summary should include the following:
   a. an explanation of the similarities and differences of solving equations and inequalities graphically and
   b. an explanation of the method for finding critical points.

I. Highlight #5 of this unit for your QUICK REFERENCE list of operations.

Solutions: **3a.** None **3b.** X = -2 **4a.** R **4b.** {X|X ≠ -2} **5d.** X = -2 or X = 3
6. {X|X ≤ -2 or X ≥ 3}, ◄———●——●——► **7.** The solutions were "opposites" of one another (one
                                          -2   3

solution set is the complement of the other).
Exercise Set: A. {X|-2 ≤ X ≤ 2.5}, ◄———●——●——►, [-2, 2.5] **B.** {X| X ≤ -4/3 or X ≥ 1}, ◄——●——●——►
                                        -2   2.5   -4/3   1

(-∞,-4/3] ∪ [1, ∞) **C.** {X|-4/3 < X < 1}, ◄———○——○——►, (-4/3, 1) **D.** R, ◄——————————►, (-∞, ∞) **E.**
                                              -4/3   1   0

{X|X ≠ -5}, ◄———○————►, (-∞, -5) ∪ (-5, ∞) **F.** Null Set **G.** {X|X ≠ 1.5}, ◄———○————►,
              -5   1.5

(-∞, 1.5) ∪ (1.5, ∞)

167

To access the FORMAT window on the TI-85, select **[F3](FORMAT)** from the second window of the GRAPH menu, then you can scroll down and turn the axes off.

You cannot turn off the axes on the Casio.

To turn the axes off, enter the PLOT OPTIONS screen and uncheck AXES.

UNIT #19
SOLVING NONLINEAR INEQUALITIES

In the previous unit we looked at solving quadratic inequalities. To solve <u>rational</u> nonlinear inequalities we will use some of the procedures from the previous unit and investigate necessary modifications.

1. Recall the methods used in the previous unit for solving quadratic inequalities. The following steps were used:

    a. The inequality was written with one side (usually the right) greater than/less than 0.

    b. The left side of the inequality was entered on the calculator at the Y1 = prompt. The standard viewing screen was used to view the graph. ([ZOOM] [6:ZStandard]).

    c. Because the right side of the inequality was equal to zero, the X-axis was used as a reference.

    d. Once the graph was displayed, the critical points were found (by using the ROOT option under the CALC menu). Recall these were the values of X that made the equation associated with the given inequality a true statement. Graphically, they were the point(s) where the graph crossed the X-axis.

    e. The critical points were graphed on a number line. They were enclosed with an open circle if the original inequality was a <u>strict</u> inequality (> or <) and by a closed circle if equality was included (≥ or ≤).

    f. If the inequality was >, the solutions were the X-values corresponding to points <u>above</u> the X-axis. Conversely, if the inequality was <, solutions were the X-values corresponding to points <u>below</u> the X-axis. Appropriate regions were shaded on the number line graph.

    The previous unit also outlined the manner in which the TABLE feature of the calculator could be used to determine appropriate solutions once the critical points were determined.

2. To solve $\dfrac{12}{X} \geq 3$ we first use a little algebra to rewrite the inequality as

    $\dfrac{12}{X} - 3 \geq 0$. (Comparing an expression to zero allows us to use the X-axis as

171

a reference point.) Enter $\dfrac{12}{X}$ - 3 at the Y1 = prompt,
press [ZOOM] [6:ZStandard]. Your display should match
the one at the right.

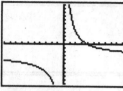

3. We need to find our critical point(s). First, circle the point where the graph
    crosses the X-axis on the above display. Now use the Root option of the
    calculator to find the critical point. X = _____

4. To confirm that 4 is a root, access your TABLE (first set the table up to begin
    at -1 and increment by 1). When X = 4, Y1 has a value of 0. This confirms
    that 4 is our root (and thus our critical point). However, notice that when
    X = 0 that Y1 = ERROR. **WHY** is there an ERROR message for this value of
    X? Go back to the original inequality, substitute 0 in for X, and answer the
    question.

    Because the fraction $\dfrac{12}{X}$ is undefined when X = 0, we must ensure that
    we DO NOT include this value for X as part of our solution. This is an
    excluded (or restricted) value.

5. a. Locate the excluded value (0) and the root (4) on the graph.

    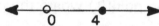

    Notice that the coordinate 0 is marked with an open circle, while the
    coordinate 4 is marked with a closed circle.
    WHY are these different?

    IT IS IMPERATIVE THAT YOU INITIALLY LOCATE YOUR EXCLUDED VALUES
    (RESTRICTED VALUES) **BEFORE** GRAPHING. Recall, these are the values that
    make <u>any</u> denominator equal 0. They must be located on the number line and
    are <u>ALWAYS</u> marked by an open circle as they are <u>NEVER</u> a part of the
    solution.

    b. Access your graph and <u>shade</u> the part(s) of the pictured number line that
    correspond to those points of the graph that are <u>above</u> the X-axis.

    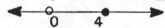

Your graph should correspond to $\{X|\ 0 < X \leq 4\}$ which is expressed as $(0,4]$ in interval notation.

c. Accessing the TABLE feature allows us to check our solution. Notice for values of X SMALLER than 0, the Y1 values are negative, and are <u>not</u> a part of the desired solution. When X is GREATER than 0 but LESS than 4 the corresponding Y1 values are positive, and thus <u>are</u> solutions to $\dfrac{12}{X} - 3 \geq 0$.

Scrolling past 4 (to X values GREATER than 4) we notice that the corresponding Y1 values are negative, and <u>not</u> a part of the desired solution.

6. Consider the inequality $\dfrac{-8}{X-4} \leq \dfrac{5}{4-X}$.

a. Rewrite the inequality so that one side equals 0: $\dfrac{-8}{X-4} - \dfrac{5}{4-X} \leq 0$

Enter the left side at the Y1 = prompt and press **[GRAPH]**. Your display should match the one at the right.

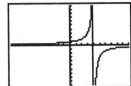

b. Locate the <u>excluded</u> value(s): X ≠ _____
(Either find them algebraically by setting denominator factors equal to 0 and solving for X or access the TABLE and search for ERROR messages in the Y1 column.)

c. We now find any critical point(s). These are the point(s) where the graph crosses the X-axis. We can use the ROOT option to try to find the root(s). Describe what happens when you try to get close to what you believe is the root.

d. To better see what is going on, we will put the calculator into **DOT** mode. To do this, press **[MODE]** and cursor down to "Connected" and then right to "Dot". Press **[ENTER]** to highlight this mode. Press **[GRAPH]**. Your display should correspond to the display at the right.

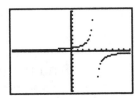

Compare this display to the one at 6a. Notice the vertical line (where we believed there was a root) is gone. In CONNECTED MODE this line connected two adjacent pixel points on the graph. However, in DOT MODE we can see that these two points are at opposite ends of the graph and should not be connected. To connect them would mean that 4 is an acceptable value for X.

The graph <u>never</u> crosses the X-axis but rather jumps from a location above the X-axis to one below it. **TRACE** and observe the X-values to confirm this.

The graph <u>never</u> crosses the X-axis but rather jumps from a location above the X-axis to one below it. **TRACE** and observe the X-values to confirm this.

e. Recall, we want the solutions to the inequality $\dfrac{-8}{X-4} - \dfrac{5}{4-X} \leq 0$. We place 4 on our number line and circle it (our excluded value) and shade to the right:

WHY?

The solution set is $\{X \mid X > 4\}$ or $(4, \infty)$ in interval notation.

f. Accessing the TABLE feature confirms the fact that values of Y1 are positive <u>below</u> X = 4 and negative <u>above</u> X = 4, thus validating our solution above.

7.   We will outline our solving method by finding the solution to the inequality $\dfrac{X}{X+9} \leq \dfrac{1}{X+1}$.

a. Rewrite the inequality with one side equal to 0. Enter the non-zero side at the Y1 = prompt. You are still in DOT MODE.

b. Find the excluded values (either by setting each denominator above equal to 0 and solving for the X values OR by accessing the TABLE and scrolling for ERROR message(s)).

X ≠ _____ and X ≠ _____

Enter these values on the number line below (at letter d.) and mark them with <u>open circles</u> so that you do not inadvertently include them in your solution).

174

c. Find the ROOTS of the equation associated with the above inequality by using the ROOT option under the CALC menu OR by solving the equation

$$\frac{X}{X+9} - \frac{1}{X+1} = 0$$ "by hand" (algebraically).

Roots: X = _____ or X = _____

Enter the roots on the number line in letter d. below and mark them with closed circles.

d. We want to shade the regions on our number line where Y1 (the graph) is below the X-axis. Access the TABLE and scroll to X = -9. Notice the Y1 values are <u>positive</u> (not part of our solution) for values of X < -9. DO NOT SHADE to the left of -9. Look at the Y1 values <u>between</u> -9 and -3. Because they are <u>negative</u> we DO SHADE this region. Continue scrolling and checking Y1 values between the values on your number line. Shade those regions in which the Y1 values are negative.

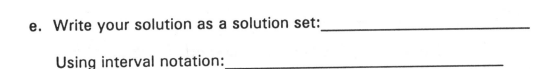

e. Write your solution as a solution set:_____

   Using interval notation:_____

f. Reset your calculator to CONNECTED MODE and compare your number line graph to the graphical display on your calculator screen. **TRACE** and observe X and Y values. Confirm that your solution above is correct. Notice that as you TRACE left that the Y values jump from negative values to positive values as the screen scrolls.

Use the steps outlined in #7 above to solve each inequality below. Begin by setting your calculator in Connected MODE with a standard viewing WINDOW. Remember to access the TABLE feature (when appropriate) and convert to Dot MODE as YOU deem appropriate. Use the combination of algebra and calculator that makes you feel comfortable.

A. $\dfrac{X^2 + 6X + 9}{X + 5} > 0$

     a. Number line graph: ⟵_____⟶

     b. Solution set: _____

     c. Interval notation: _____

B. $\dfrac{2}{X - 2} < \dfrac{3}{X}$

     a. Number line graph: ⟵_____⟶

     b. Solution set: _____

     c. Interval notation: _____

C. $\dfrac{(2X - 1)(X - 5)}{X + 3} \leq 0$

     a. Number line graph: ⟵_____⟶

     b. Solution set: _____

     c. Interval notation: _____

Note: In letter C above, did you shade only between the roots of 1/2 and 5? This problem illustrates an important point: ALWAYS look to the left (or right) of the restricted values and/or critical points to see how the graph behaves.

Set your calculator in Connected MODE. TRACE left on the graph of the above expression and go beyond the vertical line that connects the two non-adjacent pixel points. You will not see anymore graph (TRACE until your graph shifts at least once). Look at the X and Y coordinates at the bottom of the screen, and

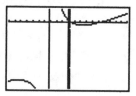

then check your window values. We will now scroll down to Ymin and enter -50. Press [GRAPH] and compare your display to the one pictured.

NOW you should understand why the region of the number line less than -3 is shaded.

8. Summarizing Results: Your summary of this unit and should include the following:
    a. what excluded values are, how to find them, and their inclusion in solutions,
    b. how to locate critical points,
    c. how to access Dot MODE and it's advantages and disadvantages in looking at solutions of non-linear rational inequalities,
    d. differences in graphs in Dot MODE and Connected MODE. Uses of the TABLE feature - as well as its limitations.

<u>Solutions:</u> **7a.** $\dfrac{X}{X+9} - \dfrac{1}{X+1} \le 0$ **7b.** $X \ne -9$ and $X \ne -1$ **7c.** $X=3$ or $X=-3$ **7d.**

7e. $\{X|-9<X\le-3$ or $-1<X\le3\}$ $(-9,-3] \cup (-1,3]$

Exercise Set: A. $\{X|-5<X<-3$ or $X>-3\}$ $(-5,-3) \cup (-3,\infty)$

B. $\{X|0<X<2$ or $X>6\}$ $(0,2) \cup (6,\infty)$

C. $\{X|X<-3$ or $\frac{1}{2}\le X\le5\}$ $(-\infty,-3) \cup [\frac{1}{2},5]$

178

Dot mode on the TI-85 is found on the **FORMAT** menu (**[GRAPH] [MORE]**) and is called **DrawDot**.

Dot mode on the Casio is accessed as follows: [SHIFT] <SET UP>, select DRAW TYPE: and press [F2](PLT) (which plots the individual points).

6.   d. To activate the "DOT" mode, enter the PLOT OPTIONS screen and uncheck CONNECT.

UNIT #20
SOLVING RADICAL EQUATIONS

This unit will investigate using the INTERSECT and ROOT options on the TI-82 to solve equations that contain radicals. Recall, the solution to an equation is the value(s) for the variable that produce a true arithmetic statement.

1.   Solve $\sqrt{3X + 7} + 2 = 7$ algebraically   Check your solution(s) by substitution:

2.   To graphically solve this equation we will first look at the graphical representation of each side of the equation.

Enter $\sqrt{3X + 7} + 2$ at Y1 and the constant 7 at Y2.

Press [ZOOM] [6:ZStandard] and compare your graph to the screen pictured at the right. Recall, we want to determine graphically where Y1 = Y2. Circle the point of intersection.

3.   Use the INTERSECT option (located under the CALC menu) to graphically find the X-coordinate of the intersection of the two graphs. Is the X-coordinate of the intersection the same as the value you found algebraically in #1? _____

If it is not, recheck your algebraic solution and your calculator solution.

4.   Now use the ROOT option to graphically solve the same radical equation. Remember, you must first rewrite the equation with all terms on one side of the equal sign and the other side equal to 0. Do this in the space below.

After clearing the expressions from Y1 and Y2, enter the non-zero side of the equation at the Y1 prompt.

5. Press [GRAPH] and compare your screen to the one pictured at the right. Circle the root, i.e. the X-intercept.

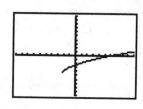

6. Use the ROOT option (located under the CALC menu) to find the root of the equation. Your root should, of course, be 6.

EXERCISE SET

Directions: Use either the ROOT or the INTERSECT option on the TI-82 to solve each radical equation below. Sketch the screen display and use the ▸Frac option (under the MATH menu) to convert all decimal answers to fractions.

A. $\sqrt{X^2 + 6X + 9} = -X + 6$

X = _____

Converted to a fraction, X = _____

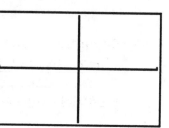

B. $\sqrt{2X + 5} = \sqrt{3 - X}$

X = _____

Converted to a fraction, X = _____

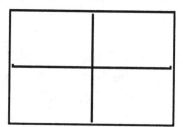

C. $\sqrt[3]{2X + 6} = 2$

X = _____

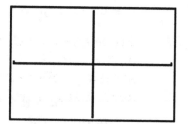

184

D. $\sqrt{X + 4} + 6 = 3$

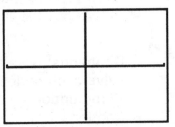

Solution: _____

If you solved the equation algebraically, the first step would be to isolate the radical. Once the radical is isolated, you should realize there are no solutions. Why?

7. Solve the equation $\sqrt{X} + 2 = \sqrt{5 - X} + 3$ algebraically in the space below. Check your solution(s) by substitution.

Algebraic Solution   Check by substitution

8. Graph the above equation by entering the left side as Y1 and the right side as Y2. Copy the display screen. How many points of intersection do you see? Use the calculator to find the solution.

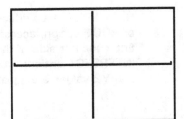

X = _____

185

Directions: Use either the ROOT or the INTERSECT option on the TI-82 to solve each radical equation below. Sketch the screen display and use the ►Frac option (under the MATH menu) to convert all decimal answers to fractions.

E. $\sqrt{2X - 3} = 3 - X$

Be Careful! Make sure BOTH the X and Y coordinates are displayed at the bottom of the screen.

X = _____

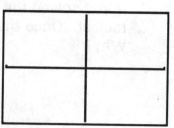

F. $\sqrt{X^2 - 12X + 36} + 5 = 7$

X = _____   X = _____

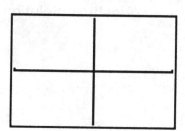

G. $\sqrt[3]{X^3 + 5X^2 + 9X + 18} = X + 2$

X = _____   X = _____

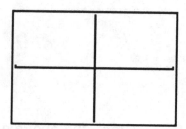

Hint: If you have difficulty finding the roots using the INTERSECT or ROOT option, access the TBL feature (set your table to begin with X = 1 and increment by 1). Enter the left side of the equation at Y1 and the right side at Y2 (as though you were using the INTERSECT option). Access your table, and scroll until you find the X-value(s) for which the Y1 and Y2 values are equal.

9. In the space below, solve $\sqrt{-2X + 6} = 3 - X$ algebraically. Check solutions by substitution.

    Algebraic solution: Check by substitution:

10. a. Use the INTERSECT option and have the calculator find the solutions. Copy your screen display.

    X = _____

    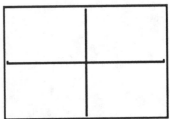

    b. Use the ROOT option and have the calculator find the solutions. Copy your screen display.

    X = _____

    c. You found two valid roots algebraically, two roots were displayed graphically, and yet only one root could be computed with the calculator. Which is the correct solution - the algebraic solution in #9 or the calculator solutions above? *Justify* your response.

d. Set your table option at a minimum of **1** and increments of **1**. Make sure you have the left side of the equation entered at Y1 = and the right side at Y2 =. Access the table. For what values of X are the Y1 and Y2 values equal?

X = _____   X = _____

This confirms your algebraic solution.

11.   **Application:** The period of a pendulum on a clock is the time required for the pendulum to complete one cycle (one "swing" from a given position back to this initial point). The formula for finding the period of a pendulum is $T = 2\pi\sqrt{\dfrac{X}{32}}$, where T is the time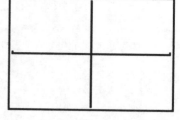
required in seconds and X is the length of the pendulum. A clock company is constructing a clock for a window display. If it takes the pendulum two seconds to complete 1 period, what is the length of the pendulum (to the nearest hundredth of a foot)?   ANS. 3.24 feet

NOTE: Work problems from your text using what you have learned in this unit. Decide what works best for <u>you</u> - algebraic solutions? the ROOT option? the INTERSECT option? Then PRACTICE.

12.   Summarizing results: On the next page, summarize what you learned in this unit. Your summary should address
      a. the use of the INTERSECT option to solve radical equations
      b. the use of the ROOT option to solve radical equations
      c. the use of the TABLE feature to find roots

Solutions: 1. X = 6 3. Yes 4. $\sqrt{3x+7} - 5 = 0$

Exercise Set: A. X = 1.5, 3/2 **B.** X = -.6666667, -2/3 **C.** 1 **D.** Null Set, Because the right side would be equal to -3 and $\sqrt{}$ is defined only for positive roots. **7.** X = 4, One is an extraneous root. **8.** X = 4 **E.** X = 2 **F.** X = 4 or X = 8 **G.** X = -5 or X = 2 **9.** X = 1 or X = 3 **10a.** X = 1 **10b.** X = 1 **10c.** {1, 3} - Both of these solutions check algebraically. **10d.** X = 1 or X = 3

UNITS * Instructors are encouraged to use Units marked by an * to <u>introduce</u> concepts.	PRE-REQ. UNIT(S)	CORRELATING CONCEPT
#21 How Does the TI-82 Actually Graph?	#1-3, #5	This unit correlates calculator graphing with hand drawn graphs.
#22 Graphing Linear Equations in Two Unknowns　　*	#10, #11, #21	Explorations of the rectangular coordinate system
#23 Preparing to Graph: Calculator Viewing Windows	#21	This unit explores pre-set viewing windows on the calculator.
#24 Where Did the Graph GO? (Adjusting the viewing window to fit your graph)	#21, #23	This unit addresses the manner in which X and Y axes can be "stretched" to obtain good graphical displays.
#25 Discovering Slope　　*	#21, #23	This exercise is designed to be done <u>before</u> slope has been formally introduced.
#26 The State Fair and Your Graphing Calculator　　*	#21, #23	An in-depth examination of one linear application: this unit may be done <u>before</u> applications are formally introduced in class.
#27 Applications of Systems of Linear Equations	#10, #11, #21, #23	Four applications of linear systems are examined in detail.
#28 Functions	#10, #21, #23	An exploration of functions: notations, domain, range, vertical and horizontal line tests
#29 Graphing Equations with Radicals　　*	#21, #23, #28	A close examination of the domain, range and graph of the square root function
#30 Discovering Parabolas　　*	#21, #23	An introduction to graphing quadratic functions (no prior quadratic graphing experience required)
#31 Exponential Functions and their Inverses	#28	The graphs and relationships between the exponential function and its inverse are examined.
#32 Predict-a-Graph	#28	Provides a collective overview of many graphs encountered by the end of Intermediate Algebra

INTRODUCTION OF KEYS

Unit Title	Keys
21: How Does the TI-82 Actually Graph?	No new keys
24: Graphing Linear Equations in Two Unknowns	DRAW 4:Vertical 1:ClrDraw
22: Preparing to Graph: Calculator Viewing Windows	ZOOM
23: Where Did the Graph Go?	No new keys
25: Discovering Slope	No new keys
26: The State Fair and Your Graphing Calculator	No new keys
27: Applications of Systems of Linear Equations	No new keys
28: Functions	DRAW 3:Horizontal
29: Graphing Equations with Radicals	No new keys
30: Disovering Parabolas	No new keys
31: Exponential Functions and Their Inverses	DRAW 8:DrawInv
32: Predict-A-Graph	No new keys

Consider the polynomial expression $\frac{3}{4}X + 6$.

1.   What is the degree of this polynomial?_____

2.   Using your graphing calculator, evaluate the polynomial $\frac{3}{4}X + 6$ for each of

the indicated values of X:

X	-8	-4	0	4	8
(3/4)X + 6					

3.   Now press [Y =] and enter (3/4)X + 6 after the "Y1 =".
Press [WINDOW] and enter the values shown at the right.
Press [GRAPH]. The line drawn represents the evaluations
of the polynomial (3/4)X + 6 for various replacement
values of X.

```
WINDOW FORMAT
Xmin=-47
Xmax=47
Xscl=10
Ymin=-31
Ymax=31
Yscl=10
```

4.   Press [TRACE] and the right arrow key until the screen display reads X = 4
and Y = 9. This means that replacing X with "4" in the polynomial

$\frac{3}{4}X + 6$ will yield a value of 9, i.e. $\frac{3}{4}X + 6 = \frac{3}{4}(4) + 6 = 9$ (see your

chart above).

5.   Use the arrow keys (left and right) to trace along the graph (as in #4) to fill in
the values of the table:

X	Y
-8	
-4	
0	
4	
8	

6. Plot each of the ordered pairs from #5 on the graph at the right. Use a ruler to connect the points to graph the line of the equation Y = (3/4)X + 6.

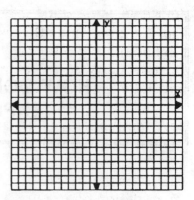

We now want to compare your hand drawn graph of the equation to the calculator drawn version. To do this, you will "draw" the graph in the same manner as the calculator.

7. Pretend the graph at the right is your calculator screen and that each box represents a pixel space. "Light up" the X-axis and Y-axis by <u>lightly</u> shading a horizontal line of boxes and a vertical line of boxes where the two axes should be located.

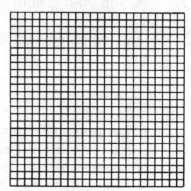

8. Now "plot" the X-intercept of (0,6) and the Y-intercept of (-8,0) by <u>darkly</u> shading the appropriate pixel space (box).

9. Place your ruler so that it connects these two pixels. Now <u>darkly</u> shade in a path of pixels along the edge of your ruler. Remember, the entire square representing the pixel must be shaded.

The three graphs below show the comparison between your hand drawn graph using the ordered pairs found in #5, the actual calculator display and the your pixel sketch.

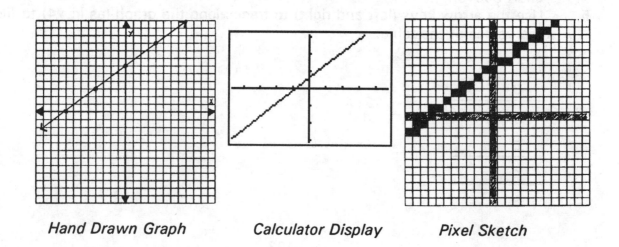

Hand Drawn Graph   *Calculator Display*   *Pixel Sketch*

194

Your calculator plots points by lighting up little squares on the screen called pixels. The TI-82 screen is 95 pixel points wide (with 94 spaces between the horizontal pixel points) by 63 pixel points high (with 62 spaces between the vertical pixel points). Because there are only a finite number of pixel spaces to light up, the calculator may only be able to "light up" a pixel that is <u>close</u> to the desired point.

Now that you have seen how the calculator must "light up" pixels to graph a straight line, we will examine what happens to curves.

10. A semi-circle with a radius of 5 units has been drawn on the graph at the right. Shade in the squares along the path of the semi-circle to simulate the calculator "lighting up" pixel spaces. (The pixels representing the X and Y axes have already been shaded for you.)

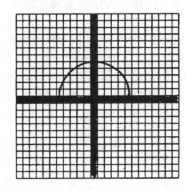

11. Now graph this semicircle on your TI-82 by entering the equation

$Y1 = \sqrt{100 - X^2}$ on the "Y=" screen. (Press [Y=] and be sure your Y1 entry looks like: $\sqrt{(100 - X^2)}$. Press [GRAPH].)

12. Sketch the graph displayed.

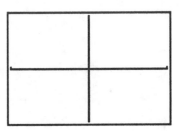

Notice how "flat" the top of the semi-circle appears. All calculator drawn lines and curves will consist of a pattern of boxes, and vertical or horizontal line segments. The **WINDOW** values you select will affect the appearance of the line or curve. The next unit examines the **WINDOW** display and how to select an appropriate viewing window.

<u>Solutions:</u> 1. first degree 2. 0, 3, 6, 9, 12 5. same as #2
12.

3. To obtain a "friendly" window for the TI-85, press [RANGE] from the GRAPH menu ([GRAPH] [F2](RANGE)) and use the values xMin = -63 and xMax = 63, with the other values the same as for the TI-82.

Recall that the TI-85 does not have a WINDOW key. Further, the size of the TI-85 window is different from that of the TI-82. The TI-85 screen is 127 pixel points wide and 63 pixel points high. This means that there are 126 pixel spaces horizontally and 62 pixel spaces vertically.

Range

Recall that the Casio does not have a WINDOW key. Instead, it has a **[Range]** key.
Press the key and enter the same values used for the TI-82.

Like the TI-82, the screen on the Casio is 95 pixels wide by 63 pixels high.

3.   In the PLOT and PLOT OPTIONS enter the values shown below.

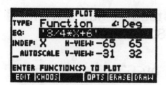 

The HP 48G screen is 131 pixel points wide and 64 pixel points high.

UNIT # 22
GRAPHING LINEAR EQUATIONS IN TWO UNKNOWNS

This unit will explore the rectangular coordinate system (also often called the Cartesian coordinate system).

1. Press [MODE] to ensure that all options on the left are highlighted. If any are not, cursor down to the item that is not highlighted and press [ENTER]. Press [Y=] and clear all entries.

2. Press [ZOOM] [4:ZDecimal]. Use your right and left cursors to move along the horizontal line (the X-axis). Observe the values of the X and Y coordinates displayed at the bottom of the screen. Record your observations below.

3. Use the cursor arrows to move the cross-hairs to the point with coordinates (0, 0), the origin. Now use your up and down cursors to move along the vertical line (the Y-axis). Record your observations about the X and Y coordinates.

4. Notice from #2 and #3 above that when the Y-coordinate is 0 you are on the X-axis and when the X-coordinate is 0 you are on the Y-axis.

5. The axes divide the screen into 4 areas (or quadrants). These are numbered using Roman Numerals, rotating counterclockwise (beginning at the top right) as I, II, III, and IV. Cursor around in each quadrant and record your observations about the signs of the X and Y coordinates in each quadrant.

Quadrant	Sign of X Coordinate	Sign of Y Coordinate
I		
II		
III		
IV		

In the units entitled "Graphical Solutions to Linear Equations" and Linear Applications" we examined the graphical solutions of linear equations in one variable. We will now look at linear equations in <u>TWO</u> variables, X and Y.

6.   Consider the linear equation X - Y = 5. To graph on a graphing calculator, the equation must first be solved for Y: Y = X - 5.

7.   There are an infinite number of ordered pair solutions to Y = X - 5. Find three of them by selecting <u>any</u> three values for X, substituting them into the equation and finding the corresponding Y values.

X	Y = X - 5	Y	Coordinates
8	Y = 8 - 5	3	(8, 3)

NO MATTER WHAT VALUE OF X YOU CHOOSE IT IS ALWAYS POSSIBLE TO FIND A CORRESPONDING Y-VALUE.

8.   Plot your points on the adjacent grid and connect with a straight line. **ALL** the points that lie on the straight line are solutions to Y = X - 5. Your three selected points were used to show the <u>path</u> that all the solutions points lie in.

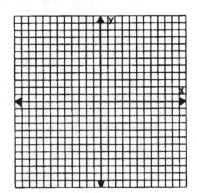

9.   a. To graph with the calculator, press [Y=], clear all entries and enter X - 5 after the Y1= prompt. Press **[ZOOM] [4:ZDecimal]** to display the graph on the ZDecimal screen. Your display should match the one pictured.

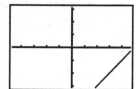

b. The TI-82 has the capability of showing you a split screen. To access this capability, press **[MODE]**, cursor down to **FullScreen**, and then right to **Split**. Press **[ENTER]** (thus choosing the Split option). Press **[GRAPH]**. Your display should match the one given.

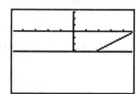

Set your TABLE to begin at **1** and to be incremented by **1**. Do this by pressing **[2nd]** **<TblSet>** **[1]**, cursor down to ᴧTbl and press **[1]** again for your increment. Access the TABLE by pressing **[2nd]** **<TABLE>**.

Scrolling up and down your TABLE displays X-values you selected (as well as an infinite quantity of other choices) and the corresponding Y-values (the value of the expression you entered at Y1 for the given X value). Stop scrolling when the X-value of 6 is displayed at the top of your table.

c. Return to the graph by pressing **[TRACE]**. Cursor right until the X and Y values displayed at the bottom of the graph screen are X = 6 and Y = 1. The blinking cursor is located at the point on the line corresponding to the top table value.

d. Press the right cursor <u>once</u>. The cursor is now located at the position where X = 6.1 and Y = 1.1. Notice there is no corresponding entry in your TABLE.

Why is there not an entry of 6.1 for X in your TABLE?

e. Reset your TABLE (press **[2nd]** **<TblSet>** to begin at 6 and to increment by 0.1 (one tenth). Access the TABLE. You should now see the desired entry. Scroll up the TABLE to X = 5.3 and Y1 = 0.3. Press **[TRACE]** and find the corresponding point on the graph.

10. What is the Y-value when X = 5.38? Neither your TABLE nor TRACING on your graph will yield the desired Y-value. You may either reset your table increment and scroll to the desired Y-value, or use the **value (Eval X)** option.

To illustrate the **value (EVAL X)** option: First, return to a Full Screen by pressing **[MODE]**, cursoring down to FullScreen, and press **[ENTER]**. Return to Full Screen viewing for the convenience of a large display. Press **[GRAPH]**. Access **value (EVAL X)** by pressing **[2nd]** **<CALC>** **[1:value]**. Enter 5.83, the value of X for which we want the corresponding Y-value. Press **[ENTER]**; the coordinates are displayed at the bottom of the screen: X = 5.83 Y = .83

Note: When using the value (EVAL X) option, the value you select for X must be between the Min and Max X-values displayed on your WINDOW screen.

11. Regardless of the number of decimal places we use for X we can <u>always</u> find a corresponding Y-value. We have 2 options for doing this:

Option #1: Incrementing our TABLE so that the desired value for Y1 is displayed for any given X-value.
Option #2: Using the **value (EVAL X)** option of the calculator.

Directions: Graph each linear equation below on the calculator. Use the ZDecimal viewing window. Sketch the screen display in the space provided. **TRACE** to find the coordinates of the X-intercept, the Y-intercept, and a third point of your choosing on the appropriate lines, then use these points to draw the line on the grid provided (using a straightedge to draw at all times). Finally, state the value of the expression (the Y1 value) for the given X-coordinate to the desired accuracy, using either the **TABLE** feature or **value** option.

A. 4X - 2Y = 12

Equation solved for Y:_____

X-intercept: _____

Y-intercept: _____

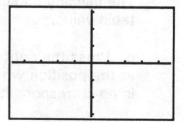

Coordinates of any third point: _____

Use the TABLE to find the Y-value when
X = -4.2. Y = _____

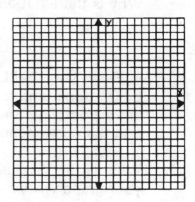

B. $Y = -\dfrac{3}{4} X - 3$

X-intercept: _____

Y-intercept: _____

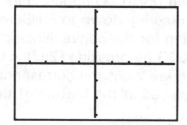

Coordinates of any third point: _____
Use **value** to find the Y-value when X = 1.7002.
Y = _____

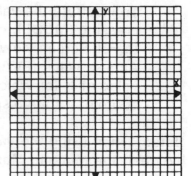

Use the TABLE to find the Y-value when X = 5.7002. Y = _____
We must use the TABLE (or change the WINDOW) when X = 5.7002 because EVAL X gives a DOMAIN error. Why won't EVAL X accept 5.7002 as a value for X?

C. Y = 2

X-intercept: _____

Y-intercept: _____

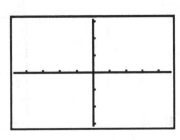
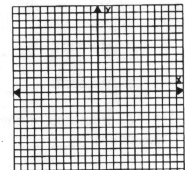

Coordinates of any third
point: _____

Use the TABLE to find the Y-value when
X = 3.695? Y = _____

A vertical line always has an equation of the form X = some constant (e.g. X = 2). If you press [Y=] to graph this line you will notice an immediate problem - it cannot be done. Your calculator is programmed to graph functions from that screen. The equation X = 2 is <u>NOT</u> a function (you will learn more about functions in the unit so entitled).

To graph X = 2, begin by returning to the Home Screen (press [2nd] <QUIT>). We will direct the calculator to <u>DRAW</u> in the line, X = 2. Press [2nd] <DRAW> and [4:Vertical]. The word "Vertical" now appears on the Home Screen. Enter [2] and press [ENTER] to instruct the calculator to draw a vertical line at X equal to 2. Because you have used the DRAW menu, you may NOT interact with this graph, i.e. TRACE, ROOT, INTERSECT, etc.

The only way to CLEAR any item from the DRAW menu is to press [2nd] <DRAW> [1:ClrDraw].

D. X = 2

X-intercept: _____

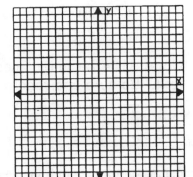

Y-intercept: _____

Coordinates of any third point: _____

E. X = -3

X-intercept: _____

Y-intercept: _____

Coordinates of any third
point: _____

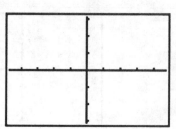 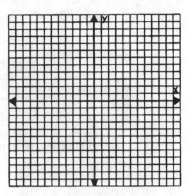

12. Summarizing Results: Summarize what you have discovered in this unit.
    Your summary should focus on the following:
    a. The graph of a linear equation in two variables
    b. The one-to-one correspondence between the coordinates of points on the
       line and the solutions to the equation
    c. The use of the TABLE feature and the value option of the calculator.

13. Highlight #9b and #10 of this unit for your QUICK REFERENCE list of
    operations.

Note: Use the techniques developed in this unit to do assigned problems in your
text. Be sure and record those techniques that are the most effective for you.

Solutions: **9d.** Because your TABLE is incremented by units (integral values of X).
Exercise Set: A. Y = 2X-6: X-int. = 3, Y-int. = -6, Y = -14.4 **B.** X-int. = -4, Y-int. = -3, Y = -4.27515,
Y = -7.275: Because the X value of 5.7002 is not between the Xmax and Xmin WINDOW values.
C. X-int. = none, Y-int. = 2, Y = 2 **D.** X-int. = 2, Y-int. = none **E.** X-int. = -3, Y-int. = none

The TI-85 does not have a TABLES feature. It does have an EVAL X option. Press **[GRAPH] [MORE] [MORE] [F1](EVAL)**.

To instruct the calculator to draw a vertical line $X = 2$, you can customize a **VERT** key on your custom menu (to use just as the TI-82 instructs) **OR** use the **ALPHA** key to type in **VERT** on the home screen. Follow **VERT** with a space before typing the 2.

NOTE: The TI-85 has a DRAW feature. It is on the second screen of the GRAPH menu. Select **[GRAPH] [MORE] [F2](DRAW) [F3](VERT)**. The calculator then shifts to the axes with a vertical line showing. You have the freedom to move this line back and forth using the left and right arrows, pressing **[ENTER]** at each place you would like to have a vertical line.

The Casio does not have a TABLES feature or an EVAL X option.

DRAW

It does have a DRAW feature. With a function graphed on the screen, press
[F3](Plot), place the cursor where you want to start your line and press [EXE]. Now
move the cursor to where you want the line to end, and press [F4](Line). The
calculator will connect these two locations with a straight line.

The HP 48G does not have the capability of showing you a split screen.

Recall that the HP 48G does not have a TABLE feature nor does it have an EVAL X option.

To draw a vertical line in the HP 48G. Do not enter an expression at EQ in the PLOT screen. Press (DRAW). Use (ZDECI) to set the Decimal Zoom values. In the main PICTURE menu, select EDIT. Move the cursor to the X value for the vertical line and move the cursor to the bottom of the screen. Press (DOT+) and move the cursor up. You can not "interact" with this graph since it is a drawn object.

SETTING UP THE GRAPH DISPLAY

You control the calculator's display through the **MODE** and **WINDOW FORMAT** screens.

1. Press the [MODE] key. **MODE** controls how numbers and graphs are displayed and interpreted. The current settings on each row should be highlighted as displayed. The blinking rectangle can be moved using the 4 **cursor** (arrow) keys. To change the setting on a particular row, move the blinking rectangle to the desired setting and press [ENTER]. We will only be concerned with rows 1,2,4,5,6 and 7.

    NOTE: Items must be highlighted to be activated.

Normal vs. Scientific notation
Floating decimal vs. Fixed to 9 places
Type of angle measurement
Type of graphing (Func = function)
Graphed points connected or dotted
Functions graphed one by one
Screen can be split to view two screens
   simultaneously

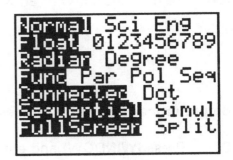

2. To return to the home screen, at this point, press [CLEAR] or [2nd] <QUIT>.

3. Press [Y =]. The calculator can graph up to 10 different equations at the same time. Because we left **MODE** in the sequential setting, the graphs will be displayed sequentially. Note that you must cursor down past Y8 to see (and hence access) Y9 and Y10. The display of the equations you enter on the "Y =" screen is controlled by the size of the viewing rectangle. You determine the dimensions of the viewing rectangle by the values you enter on the **WINDOW** screen.

4. Press [ZOOM] [6:ZStandard] [WINDOW]. You should see the screen at the right. We call this the standard viewing window. The information on this screen tells you that with a rectangular coordinate system the X-values will range from -10 to 10 and the Y-values will range from -10 to 10. The interval notation for this is [-10,10] by [-10,10]. The Xscl = 1 and Yscl = 1 settings indicate that the tic marks on the axes are one unit apart. You can change the values entered on this screen by using the cursor arrows to move to the desired line and typing over the existing entry. When you draw a graph, you can set the desired viewing rectangle on the calculator as well as scale the X-axis and Y-axis.

5. Be sure the cursor is on the word **WINDOW** and use the right arrow key to cursor over to **FORMAT**. Press [ENTER]. The following settings should be highlighted:

Graphs on a rectangular coordinate system
Cursor location is displayed on screen
Graphing grid is not displayed
Axes are visible
Axes are not labelled with an X and Y

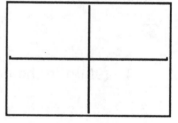

Before proceeding further, press [Y=] and clear all entries.

6. Press [**WINDOW**] and enter Xmin = -5, Xmax = 5, Xscl = 1, Ymin = -12, Ymax = 7, Yscl = 1. Be sure to use the gray [(-)] key for the minus signs. Press [**GRAPH**] to view the coordinate axes. Count the "tic" marks on the axes and see how these marks correspond to the max and min values you set. Label the last tic mark on each axis (i.e. farthest tic mark left, right, up and down) with the appropriate integral value.

7. Now change the viewing window to Xmin = -20, Xmax = 70, Xscl = 10, Ymin = -5, Ymax = 15, Yscl = 3 and again press [**GRAPH**] to view the axes. How many tic marks are on the positive portion of the X-axis?_____ How many units does each of the tic marks on this axis represent?_____ Based on your last two answers, how many units long is the positive portion of the X-axis?_____ Does this number correspond to the Xmax value given in the problem? _____ In your own words, explain what is happening.

8. To help you "de-bug" errors in your graphing set ups later on, describe what you think would happen if you set Xmin = 10 and Xmax = -5. (You might want to draw your own set of coordinate axes and try to label them in this manner.) Now do this and press [GRAPH]. What did happen?

9. Reset Xmin = -10, Xmax = 10 and describe what you think will happen if you set Ymin = 5, Ymax = 5. Again, press [GRAPH]. What did happen? (Try drawing your own set of axes and labeling them as indicated.)

10. What should the relationship between Max and Min be? (i.e. Min > Max, Min < Max, or Min = Max)

11. Reset your viewing window to ZStandard, by pressing [ZOOM] [6:ZStandard].

ENTERING EXPRESSIONS TO BE GRAPHED: THE [Y=] KEY

12. Press [Y=]. On the screen "Y1=" is followed by a blinking cursor. Anything else can be cleared by pressing [CLEAR]. Enter Y = -2X + 6, and press [ENTER]. The cursor is now on the second line following "Y2=". At this prompt, enter the equation Y = (1/2)X - 4. Note that the equal signs beside both Y1 and Y2 are highlighted. This means that both equations will be graphed. Press [GRAPH] to display the graph screen. See display at the right.

Note: Y = (1/2)X - 4 would be displayed in your text as $y = \frac{1}{2}x - 4$.

GRAPHING

13. To graph Y = -2X + 6 only, press [Y=] and use the arrow key to move the cursor over the equal sign beside Y2. Press [ENTER]. Notice that the equal sign beside Y2 is *not* highlighted, whereas the equal sign beside Y1 *is* highlighted. Press [GRAPH] - only the highlighted equation, Y1, is graphed.

14. On the viewing screen at right, the calculator draws a set of axes whose minimum and maximum values and scale match your choices under **WINDOW**. The graph of Y1 is drawn from left to right. Return to the "Y =" menu by pressing **[Y =]**. Cursor down to Y2 and "turn on" this graph by highlighting the equal sign. Press **[GRAPH]** and notice that the two graphs are drawn in sequence. (We chose **SEQUENTIAL** from the **MODE** menu earlier.)

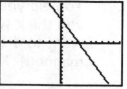

ALTERING THE VIEWING WINDOW

The last unit addressed the size of the calculator screen (viewing window). Because the screen is 95 pixel points wide by 63 pixel points high there are 94 horizontal spaces and 62 vertical spaces to light up. When you TRACE on your graph, the readout changes according to the size of the space. You can control the size of the space by the following formulas:

$$\frac{Xmax - Xmin}{94} = \text{horizontal space width,} \quad \frac{Ymax - Ymin}{62} = \text{vertical space height.}$$

We will now examine some preset viewing windows and how they affect the pixel space size.

15. Press **[ZOOM]**. There are nine entries on this screen. The down arrow key can be used to view entries 8 and 9.

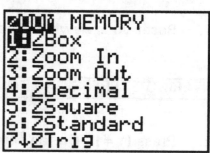

Boxes in and enlarges a designated area.
Acts like a telephoto lens and "zooms in".
Acts like a wide-angle lens and "zooms out".
Cursor moves are ONE tenth of a unit per move.
"Squares up" the previously used viewing window.
Sets axes to [-10,10] by [-10,10].
Used for graphing trigonometric functions.

Cursor moves are ONE integer unit per move.
Used when graphing statistics.

16. a. **ZDecimal** is useful for graphs that require the use of the calculator's TRACE feature. If you apply the horizontal space width formula, $\frac{Xmax-Xmin}{94} = \frac{4.7-(-4.7)}{94} = 0.1$, you discover that the X-values change by one-tenth of a unit each time you move the cursor. This is why we say that the ZDecimal screen yields "friendly" values when TRACING. (In general, Xmax - Xmin needs to be a multiple of 94 to produce a "friendly" screen.)

212

b. **ZInteger** is useful for application problems where the X-value is valid only if represented as an integer (such as when X = number of tickets sold, number of passengers in a vehicle, etc.). If you apply the horizontal space width formula, $\frac{Xmax-Xmin}{94} = \frac{47-(-47)}{94} = 1$, you discover that the X-values change by one unit each time you move the cursor.

```
WINDOW FORMAT
Xmin=-47
Xmax=47
Xscl=10
Ymin=-31
Ymax=31
Yscl=10
```

c. **ZStandard** provides a good visual comparison between your own hand sketched graphs (or textbook graphs) that are approximately [-10,10] by [-10,10]. Applying the horizontal space width formula,

$\frac{Xmax-Xmin}{94} = \frac{10-(-10)}{94} \approx 0.212765974$, you discover

```
WINDOW FORMAT
Xmin=-10
Xmax=10
Xscl=1
Ymin=-10
Ymax=10
Yscl=1
```

that the X-values will change by .212765974 each time you move the TRACE cursor. If you are using the TRACE feature, you will usually want a screen with "friendlier" X-values than this one provides.

17. ZDecimal frequently does not provide a large enough viewing window. When this is the case, you may multiply the Xmin and Xmax by the same constant and the Ymin and Ymax by the same constant to produce a larger viewing rectangle which still provides cursor moves in tenths of units. Multiplying Xmin and Xmax by 2 would mean cursor moves of two-tenths of a unit: $\frac{Xmax-Xmin}{94} = \frac{2(4.7)-2(-4.7)}{94} = 0.2$,

```
WINDOW FORMAT
Xmin=-9.4
Xmax=9.4
Xscl=1
Ymin=-6.2
Ymax=6.2
Yscl=1
```

whereas multiplying by 3 would mean cursor moves of three-tenths of a unit. The screen at the right is the ZDecimal screen with the max and min values multiplied by 2. This WINDOW will be referred to in the future as ZDecimal x 2.

18. Press [Y=] and "turn off" Y2 = (1/2)X - 4 (as described in #13), and "turn on" Y1 = -2X + 6. Press [ZOOM] [4] to view the graph in the ZDecimal screen. The screen is shown at the right. Press [WINDOW] and increase the max and min values by a factor of 2 (ZDecimal x 2). The

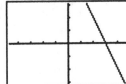

range values for X will now be [-9.4,9.4] and the y values will be [-6.2,6.2]. Pressing [GRAPH] will produce the screen at the left. You **do not** have to use the same factor for both the Xmin/Xmax and Ymin/Ymax. However, since X is the independent variable the cursor moves will be determined by the factor chosen for the Xmax and Xmin values.

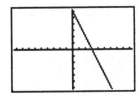

19. Press "**Y=**" and clear all entries. Enter Y1 = $\frac{2}{3}X$ + 6 and

Y2 = $-\frac{3}{2}X$ - 5 . (Did you remember to put parenthesis around the fractions?)

20. $Y1 = \frac{2}{3}X + 6$ and $Y2 = -\frac{3}{2}X - 5$ are perpendicular lines whose intersection forms a 90° angle. Press [ZOOM] [6] for the standard viewing rectangle. Notice that the lines do not appear to be perpendicular. This is because the screen is rectangular - not square. Sketch the graph display <u>exactly</u> as it appears on your calculator screen.

21. Press [ZOOM] again and this time select [5:ZSquare]. ZSquare "squares up" the viewing screen based on the previous viewing window. The lines should now appear to be perpendicular. Sketch the graph display as it now appears. Notice that the tic marks are all evenly spaced.

22. Press [WINDOW] to see how the Max and Min values were affected. Enter the WINDOW values displayed. Explain how the viewing window is different from the ZStandard viewing window.

```
WINDOW FORMAT
 Xmin=
 Xmax=
 Xscl=
 Ymin=
 Ymax=
 Yscl=
```

NOTE: The ZDecimal screen (or any multiplicity of this screen) will provide you with a "squared up" graph.

23. CLEAR all entries on the Y= screen. Using $Y1 = \frac{1}{2}X - 4$, press [ZOOM] [4:ZDecimal] and at the right, sketch the screen as displayed- including the "tic" marks on the axes.

24. Now, press [ZOOM] [8:ZInteger] (pause for your graph to be displayed) [ENTER] and at the right, sketch the screen as displayed- including the "tic" marks on the axes.
NOTE: ZDecimal and ZStandard do not require you to press [ENTER] to activate the viewing WINDOW, but ZInteger demands it.

25. In your own words, explain the differences in the displays in #23 and #24. What accounts for these differences.

26. Each of these graphs (in #23 and #24) represent the same equation. Based on the indicated scale values, determine the indicated maximum and minimum values for each of the axes in problems 23 and 24.

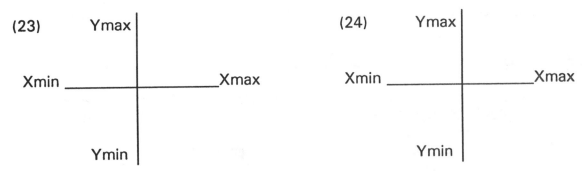

(23) Ymax, Xmin, Xmax, Ymin

(24) Ymax, Xmin, Xmax, Ymin

27. Highlight the **screens** on #1 and #5 for your QUICK REFERENCE list of operations. These are the default settings. They will remain fixed unless you change them (or someone who has used your calculator changes them).

Solutions: 6. left:-5, right:5, top:7, bottom:-12   7. 7,10,70,yes

8.   9.   10. Max > Min   20.

21.

22. To have a "square" screen, tic marks must be evenly spaced. This was accomplished by adding tic marks to the X-axis.

23.   24.

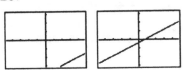

25. The difference was the amount and position of graph displayed. More of the graph was displayed on the ZInteger screen. This was because the **scales** were different on the two screens.

26. (23) Xmin: -4.7, Xmax: 4.7, Ymax: 3.1, Ymin: -3.1
    (24) Xmin: -47, Xmax: 47, Ymax: 31, Ymin: -31

UNIT #23 APPENDIX
TI-85

SETTING UP THE GRAPH DISPLAY

The TI-85 does not have the same MODE screen as the TI-82. However, all the options available on the TI-82 MODE screen are available on the TI-85 on one of the following two screens.

The screen on the left is the [2nd] <MODE> screen on the TI-85. The screen on the right is accessed by pressing [GRAPH] [MORE] [F3](FORMT)

The TI-85 can graph up to 99 equations.

GRAPHING

You cannot place the cursor over the equal sign in the [y(x)=] screen of the TI-85. Instead, to select a function to be graphed, from the [y(x)=] screen, press [F5] (SELCT) and the equal sign will be highlighted.

ALTERING THE VIEWING WINDOW

15.   Press [GRAPH] [F3](ZOOM). Pressing [MORE] will display all the ZOOM menu selections for the TI-85.

16.   These are ZDecimal and ZInteger windows for the TI-85. The values in the ZStandard window are the same.

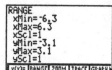   

ZDecimal   ZInteger

The formula for determining the pixel space width and height for the TI-85 are $\dfrac{Xmax - Xmin}{126}$ = horizontal space width and $\dfrac{Ymax - Ymin}{62}$ = vertical space height. In general, Xmax - Xmin needs to be a multiple of 126 to produce a friendly screen.

216

```
┌─────────────────────────────────────────────────────┐
│                 UNIT #23  APPENDIX                    │
│                  Casio fx-7700GE                      │
└─────────────────────────────────────────────────────┘
```

SETTING UP THE GRAPH DISPLAY

The Casio does not have the same MODE screen as the TI-82. However, most of the options available on the TI-82 MODE screen are available on the Casio on either of the following two screens.

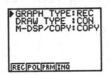   

The screen on the left is the **[SHIFT] <SET UP>** screen on the Casio. (Note: Your calculator should be set up with M-DSP/COPY:M-DP) The screen on the right is the **[SHIFT] <DISP>** screen.

You cannot place the cursor over the equal sign in the **GRAPH** screen of the Casio. Instead, to select a function to be graphed, from the **GRAPH** screen, press **[F5](SEL)**, place the cursor on the function you would like to graph and press **[F1](SET)**. To cause a function *not* to be graphed, you place the cursor over it and press **[F2](CAN))**.

ALTERING THE VIEWING WINDOW

The Casio does not have a ZDecimal or a ZInteger option, but this can be achieved by using the following settings in the Range window.

ZDecimal: ZInteger:

Note: The settings for ZDecimal above can be quickly obtained by pressing **[Range]** **[F1](INIT)**.

> **Example:** Set a screen which is "friendly" and which is about 10 units on each side of zero.
> **Keystrokes:** From the **GRAPH** screen, select a function. Press **[Range]** **[F1](INIT) [EXIT]** then **[F6](DRW)**. Now press **[SHIFT]** **<ZOOM> [F4](x 1/f)**.

You can maintain a "friendly" window by simply selecting the values of ZDecimal and then using the **(x f)** and **(x 1/f)** options from the zoom menu.

SETTING UP THE GRAPH DISPLAY

You control the calculator's display through the CALCULATOR MODES and the PLOT/PLOT OPTIONS screens. Press [→] <MODES> to enter the CALCULATOR MODES screen. To change a setting, move the cursor onto the setting and press (CHOOS). Then use the arrow keys to select the setting you want and press [ENTER].

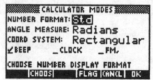

To return to the home screen, press [ON] or (OK).

Expressions to be graphed must be entered and/or selected in the PLOT application. Recall that you can define variables for each expression and use the CHOOS menu command. Graphs are displayed sequentially because SIMULT is left unchecked in the PLOT OPTIONS screen.

Recall that you must set the standard viewing window for the HP 48G. (See Unit#10 Appendix.) Use the following settings:

ALTERING THE VIEWING WINDOW

In the PICTURE environment the ZOOM menu key activates the ZOOM menu. There are 15 ZOOM menu options. These are summarized in Chapter 22 of the User's Guide. BOXZ corresponds to ZBox of the TI-82, ZIN to Zoom In, ZOUT to Zoom Out, ZDECI to ZDecimal, ZSQR to ZSquare, ZTRIG to ZTrig, and ZINTG to ZInteger. There are no menu options corresponding to ZStandard and ZoomStat. One difference between the HP 48G's and TI-82's Decimal and Integer Zooms is that the HP 48G does not rescale the vertical axis. In order to mimic the TI-82, you must set the vertical range from - 3.1 to 3.2 for the Decimal Zoom, and from -31 to 32 for the Integer Zoom. Under ZDECI, the horizontal range is from -6.5 to 6.5. Under ZINTG, the horizontal range is from -65 to 65.

Many times students are frustrated when the equation they have carefully keystroked into the "Y=" screen does not appear when **GRAPH** is pressed. What actually happens to the graph? Suppose you graphed $Y = 2X^2 + 4X + 12$ on your graph paper and when you graphed this same equation on your calculator you set the viewing window to ZStandard. The figure at the right indicates the handsketched graph with the section displayed on the ZStandard screen outlined in a bold black line. The viewing window selected is not large enough to display the graph. This exercise will give you the practice necessary to feel confident about setting the visual display you see on the calculator's graph screen.

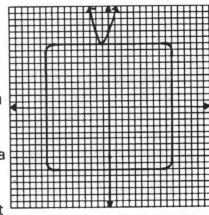

1. Before proceeding, press [ZOOM] [6:ZStandard] to set the standard viewing WINDOW. Enter Y1 = 4X - 18 on the "Y=" screen and press [GRAPH]. The graph at the right should be displayed. The X-intercept is visible, but the Y-intercept is not.

2. If the equation you have entered is not displayed on your graphing screen, the first item to be checked is the entry of the equation on the "Y=" screen. Is the equation *SELECTED* to be graphed? That is, is the equal symbol highlighted? If it is, proceed. If it is not, move the cursor over the equal sign, and press [ENTER] to highlight the equal sign, to activate the equation.

3. If the equation is activated, then you begin the process of adjusting the WINDOW by locating the X and Y-intercepts of the graph.

4. Press [TRACE] to activate the TRACE cursor. At the bottom of the screen you should see the cursor's location of X = 0 and Y = -18 displayed. This is the Y-intercept.

5. Exit the GRAPH screen and enter the WINDOW screen by pressing [WINDOW]. Use the down arrow key to move the cursor to Ymin and replace the Ymin with a value smaller than -18. Selecting a value smaller than -18 allows for a clear view of the Y-intercept.

6. Ymin = -20 was selected. Pressing **[GRAPH]** now displays the screen at the right. This a satisfactory graph because it displays both the X and Y-intercepts.

7. Reset the WINDOW to ZStandard (press **[ZOOM]** **[6:ZStandard]**).

8. Now try an equation whose graph may not be familiar. Press **[Y=]** and enter Y1 = X^3 - 15X^2 + 26X and press **[GRAPH]**. A satisfactory graph of this equation should display all of its interesting features.

9. Activate **[TRACE]**; TRACE along the graph to the right (using the right arrow key) and record the X and Y-intercepts as you encounter them. Remember, you are not on the ZInteger or ZDecimal screens. Your X-intercepts may only be close approximations with the TRACE feature. (HINT: This is a third degree equation. There could be three X-intercepts.) Each of the screens below indicate the points closest to the X-intercepts that you should be able to locate.

X and Y-intercept   X-intercept ≈ 2   X-intercept ≈ 13

10. Do you see all the peaks (maximums) and valleys (minimums) of the graph? A satisfactory viewing window is a window that includes the X and Y-intercepts as well as all the peaks and valleys of the graph. TRACE the curve again, going left this time, to determine the lowest **Y value** (valley/minimum) and the highest **Y value** (peak/maximum) displayed (to the nearest whole number value).

valley/minimum = _____   peak/maximum = _____

11. Press **[WINDOW]** and adjust your Max and Min values for both X and Y to include the intercept points on the X-axis and the maximum and minimum Y values. It is suggested that you enter values that are a few units larger (or smaller) than the intercepts, maximum, and minimum recorded above in #9 and #10. Enter your screen values here:

Xmin = ____   Ymin = ____
Xmax = ____   Ymax = ____

12. Sketch the graph displayed using the WINDOW you determined in #11.

13. Follow the directions outlined in #9 and #10 and set a good viewing window for $Y = X^2 + 2X - 99$. Enter the necessary values to determine a good WINDOW.
Y-intercept = _____
X-intercepts = _____, and _____
Ymin = _____

Note: The <u>WINDOW values</u> that you use need to be a few units larger (or smaller) than the intercepts and minimum recorded above.

14. Enter your WINDOW values on the screen and sketch the graph that is displayed for these values.

15. Your graph should look *similar* to the one displayed below. If your WINDOW values are different from the WINDOW screen that is displayed, then your graph may look slightly different. It is important that you are able to see the X and Y-intercepts **and** the full curve (i.e. all peaks and valleys). If you can see all of these, then you have a satisfactory graph.

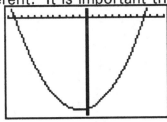

```
WINDOW FORMAT
Xmin=-12
Xmax=12
Xscl=1
Ymin=-105
Ymax=10
Yscl=1
```

EXERCISE SET

Directions: Begin each problem by viewing the graph in the ZStandard viewing window. Graph each of the equations by setting a good viewing window.
i. Record the values you used to determine your WINDOW.
ii. Sketch the graph that is displayed for your new WINDOW values.
iii. Record the WINDOW in interval notation.

A. $Y = -2X^2 + 19$

Y-intercept = _____   Ymax = _____

X-intercepts = _____, and _____

[_____, _____] by [_____, _____]
  Xmin   Xmax   Ymin   Ymax

B. $Y = (1/2)X^4 - 10X^2 + 25$

Y-intercept = _____

X-intercepts = _____, _____, _____,

_____

Ymax = _____

Ymin = _____, _____

[_____,_____] by [_____,_____]
  Xmin   Xmax   Ymin   Ymax

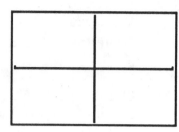

C. $Y = \sqrt[3]{X^2 - 5X - 300}$

Y-intercept = _____

X-intercepts = _____, _____,

Ymin = _____

[_____,_____] by [_____,_____]
  Xmin   Xmax   Ymin   Ymax

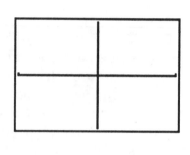

16.   Summarizing Results: In you own words, summarize what values are necessary to setting a good viewing WINDOW and carefully explain why.

Slope is a measure of the inclination of a line. The following unit will investigate this concept.

1. Use the TI-82 to graph the linear equations below. Use the WINDOW values pictured.

   In the table, record at least 5 values (in increasing order) for X and determine the corresponding Y value. (Hint: set the TABLE on your calculator to begin with X=0 and to increment by 1, then record your X and Y values in the given table). You can also use the TRACE feature. In either case, record only integral values for X and Y.

   Graph the line on the grid provided. USE A STRAIGHT EDGE TO DRAW ALL LINES.

   Answer the questions printed beneath each grid.

   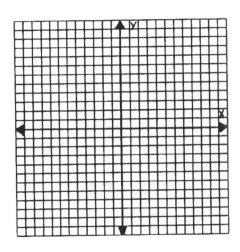

   a. $Y = \dfrac{3}{5}X$

   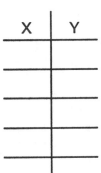

X	Y

   Look at the increments of Y recorded in your chart. As you read the successive Y values, by what constant do they increase? _____

   Look at the increments of X recorded in your chart. As you read the successive X values, by what constant do they increase? _____

   Record the RATIO of the constant Y increment to the constant X increment: _____   (ANS. 3/5)

b. Y = X

X	Y

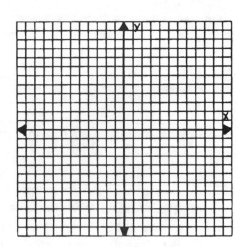

Record the Y increment: _____

Record the X increment: _____
Record the RATIO of the constant Y increment to the constant X increment:_____

c. Y = 2X

X	Y

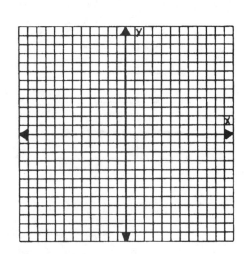

Record the Y increment: _____

Record the X increment: _____

Record the RATIO of the constant Y increment to the constant X increment: _____

d. $Y = \frac{7}{2}X$

X	Y

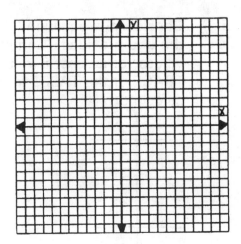

Record the Y increment: _____

Record the X increment: _____

Record the RATIO of the constant Y increment to the constant X increment: _____

2. In the graphs a-d above, compare the ratio of the Y increment to the X increment with the coefficient of X in the original equation. What do you notice?

3. What do you notice about the inclination (slope) of the line as the X-values increase?

4. Consider the line with equation $Y = 10X$. Would it be steeper than the lines previously graphed or shallower?

Check your response by graphing it on the TI-82 and comparing it with the previous 4 graphs. Sketch your graph display.

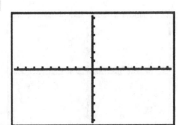

5. Consider the equation Y = -X. Graph it on the grid provided. Complete the table of values for at least 5 values of X (use either TRACE or the TABLE feature of the calculator).

X	Y

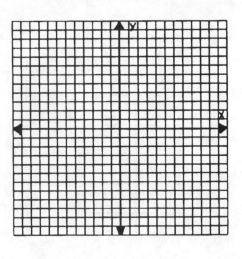

Ratio of change in Y over change in X = _____

This ratio of change (slope) is always the same as the coefficient of X. Slope is a comparison of vertical change to horizontal change.

6. Compare the line you graphed in #5 to the lines graphed in #1. What effect does the negative coefficient of X have on the line? Hint: Look at the way the lines either RISE or FALL as you scan them from left to right.

7. Graph each line below (make sure your WINDOW values are the same as in #1). Once you have entered the equation, press TRACE and record the coordinates displayed at the bottom of the screen.

a. Y = X + 2

X = ___ Y = ___

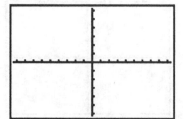

b. Y = X + 4

X = ___ Y = ___

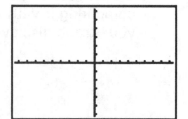

c. Y = X - 2

X = ___ Y = ___

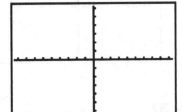

d. Y = X - 5

X = ___ Y = ___

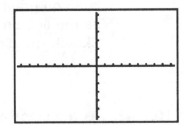

8. The coordinates displayed are the Y-intercepts. Compare the Y-coordinate in the ordered pair with the constant in the equation. What do you notice?

9. Consider the linear equation $Y = \dfrac{2}{3}X + 2$. **WITHOUT** using the calculator, answer each of the following questions.

a. What is the Y-intercept? _____

b. What is the slope of the line? _____

c. This means for every vertical change of _____ units there will be a horizontal change of _____ units.

d. Graph the line on the grid below by first graphing the Y-intercept.

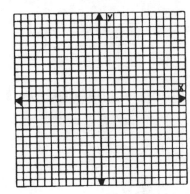

Now use the slope to locate another point. Remember slope is the ratio of $\dfrac{Change\ in\ Y}{Change\ in\ X}$; count up 2 units from the Y-intercept and right 3 units. Plot that point. Using a straightedge, draw a line through the point you just located and the Y-intercept. Record the integral values (used in your graph) in the table below:

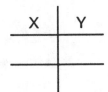

X	Y

e. Enter the above equation on the TI-82 (make sure you enclose the fraction in parentheses). Either access the TABLE or TRACE to ensure the line you graphed above in letter "d" is the same as the line on your calculator screen. Sketch the display.

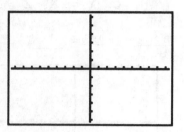

10. What is the slope of the line with equation Y = 2X + 3? _____

What is the slope of the line with equation Y = 2X - 4? _____

Graph both lines on the TI-82. Notice they are parallel. What is true about the slopes of parallel lines?

11. What is the slope of the line with equation $Y = \frac{2}{3}X + 2$? _____

What is the slope of the line with equation $Y = -\frac{3}{2}X - 1$? _____

Graph both lines on the TI-82. Notice they are perpendicular, i.e. meet at a right angle. What is true about the slopes of perpendicular lines?

12. **Application:** Apples cost $.40 per pound.
    a. Fill in the chart below that relates the total cost to the number of pounds purchased.

Pounds Purchased	Total Cost
1	
2	
3	
4	
5	

b. Write an equation to represent the relationship between pounds and cost (where Y = cost and X = pounds purchased).

c. Should you expect the graph of this line to rise or fall as you scan from left to right? _____

What information from the equation did you use to base your response on?

d. Graph the equation, Y = .40X on the TI-82 and sketch the display.
Was your prediction about the slope accurate?

e. Press [TRACE] and trace along the graph until a cost of $6.00 is displayed. Sketch the graph display at this point.

What information does the X value tell you?

13. Summarizing Results: Summarize what you have learned about the equation of a line in Y = mX + b form. Include the following points:
a. Which value in the equation Y = mX + b effects the slant (slope) of the line, and how?
b. Which value indicates the Y-intercept?
c. How do slopes of parallel lines compare?
d. How do slopes of perpendicular lines compare?

Solutions: **1a.** Y-inc.: 3, X-inc.: 5, ratio: 3/5 **1b.** Y-inc.: 1, X-inc.: 1, ratio: 1/1 **1c.** Y-inc.: 2, X-inc.: 1, ratio: 2/1 **1d.** Y-inc.: 7, X-inc.: 2, ratio: 7/2 **2.** They are the same. **3.** The line becomes steeper. **4.** Steeper **5.** Ratio: -1/1 **6.** When the coefficient of X is positive, the line rises when viewed from left to right. When the coefficient of X is negative, the line falls when viewed from left to right. **7a.** X=0, Y=2 **7b.** X=0, Y=4 **7c.** X=0, Y=-2 **7d.** X=0, Y=-5 **8.** They are the same. **9a.** 2 **9b.** 2/3 **9c.** 2, 3 **10.** 2, 2, The slopes are equal. **11.** 2/3, -3/2, The slopes are negative reciprocals of one another. **12a.** Cost column: $.40, $.80, $1.20, $1.60, $2.00 **12b.** Y=.40X **12c.** Rise, The sign of the coefficient of X. **12d.** Yes **12e.** The number of pounds purchased.

Since the TI-85 does not have a TABLES feature, you must rely on the TRACE feature to fill in the table of values. Begin TRACING at $X = 0$ and record only integral values for X and Y.

Since the Casio does not have a TABLES option, you must rely on the TRACE feature to fill in the table of values. Begin TRACING at X = 0 and record only integral values for X and Y.

UNIT #25 APPENDIX
HP 48G

1. Set the horizontal range from -9.4 to 9.4 and the vertical range from -6.2 to 6.4.

Note that under the FCN menu found in the PICTURE environment, there is a command called SLOPE that will compute the slope of the function at the X-value of the cursor.

The state fair is coming to your hometown for 3 days. This unit will use the calculator to explore cost as related to rides at the fair.

1. The first day of the fair, admission is $5, and it costs $.75 to ride each ride. We can represent this algebraically with the equation
    $Y = 5 + .75X$ where X represents the number of rides and Y represents the total amount of money spent.
    Enter the equation at Y1=, press **[ZOOM] [6]** (to access the standard viewing screen). Press **[TRACE]**.
    Record the values you see: X = _____   Y = _____
    Sketch the screen displayed:
    How would you interpret this information in
    the context of the state fair?

2. Press your right cursor (**[▶]**) and observe the X and Y values displayed at the bottom of the screen. These decimal representations are meaningless in the context of our problem. We need <u>integers</u> for X. Why?

    Press **[ZOOM] [8]**. <u>Wait</u> for your graph to appear. Use the **[◀] [▼]** to return to X = 0 and Y = 0, then press **[ENTER]**. We are accessing the ZInteger viewing screen.

3. Press **[TRACE]**, cursor right and observe the X and Y values displayed at the bottom of the screen. Answer each question below:

    a. How much would you spend if you rode 5 rides? _____

    b. How much would you spend if you rode 15 rides? _____

    c. If you cursor left past the Y-axis, you will get negative values of X and Y. These points are meaningless in the context of our problem. Why?

235

4. The normal price of the fair is $1 admission and $1.25 per ride. The second night the fair is in town it will run its normal rates. Write the equation that would represent cost in terms of number of rides in the space below. (Hint: You may use the equation from #1 as a model.)

5. Enter the equation from #4 (Y = 1 + 1.25X) at the Y2 = prompt on your calculator. Before pressing **[GRAPH]**, turn off the graph of Y1 by placing the cursor over the = sign beside Y1 and pressing **[ENTER]**. Now press **[GRAPH]** and you will see the graph of the equation entered at Y2 displayed. (Sketch the screen.)
Use **[TRACE]** to answer the questions below:

   a. How much would you spend if you rode 5 rides?_____

   b. How much would you spend if you rode 10 rides?_____

6. "Turn on" the graph of Y1 by locating the cursor over the = sign beside Y1 and pressing **[ENTER]**. Press **[GRAPH]** and observe the graphs representing the cost and number of rides for both nights.

   a. How many rides would you have to ride for the cost to be the same no matter which night you attended the fair? _____

   b. Notice the difference in slopes of the two lines (line Y2 is steeper than line Y1). What does this mean in the context of our problem?

7. The third night that the fair is in town, the managers decide to run a special: $10 admission and no additional cost per ride. We would represent this with the equation Y = 10. Enter this equation at Y3. Sketch the display when all three equations are graphed.

   a. How many rides would you have to ride for the $10 price to be a better deal than the first night? _____
   Hint: You might want to turn the graphs ON and OFF to make comparisons.

b. How many rides would you have to ride for this to be a better deal than the second night? _____

c. In considering which night would be the best night to attend the state fair we have only considered attendance based on cost. What other factors would have to be considered before making an informed decision?

d. Which night do you believe would have the largest crowd? Why?

Solutions: 1. $X=0$, $Y=5$, If you rode no rides, it would still cost you five dollars. **2.** Because X represents the number of rides, and you cannot ride a decimal part of a ride. **3a.** $8.75 **3b.** $16.25 **3c.** It is impossible to ride a negative number of rides. **4.** $Y=1+1.25X$ **5a.** $7.25 **5b.** $13.50 **6a.** 8 **6b.** The cost increases more rapidly at the rates imposed the second night. **7a.** 7 **7b.** 8 **7c.** Answers will vary. **7d.** Answers will vary.

238

The graphing calculator enables you to solve systems of linear equations in two variables quite easily. The following application problems illustrate the use of systems in basic decision making.

1.   The local pool offers a savings plan at the beginning of the summer. In this plan, an individual can purchase a pass for $50. This allows that individual to utilize the facilities as often as he/she wishes. Without the pool pass each visit costs $4.

   a. Write 2 equations, one in which Y represents the cost of the pool pass and the other with Y representing cost and X representing the number of times the pool is visited.

   Equation 1: Y = _____

   Equation 2: Y = _____

   b. Enter these equations at the Y1 = and Y2 = prompts. Ensure you are in the standard viewing screen by pressing [ZOOM] [6:ZStandard]. Adjust the viewing rectangle until the point of intersection for the 2 graphs is clearly visible. (HINT: Try the ZOOM OUT option or other methods for adjusting your viewing window that have been discussed previously.) Describe in detail what you did to obtain a suitable display of the graphs. (Recall that the number at the top right of your display screen indicates the graph you are on once you access the TRACE feature. The #1 corresponds to the equation at the Y1 prompt and the #2 corresponds to the equation at the Y2 prompt).

   c. Sketch your display and circle the point of intersection. Use the INTERSECT option of the CALC menu to obtain the X and Y coordinates of the intersection point.

   X = _____   Y = _____

d. How many times must an individual visit the pool for the pass to be the best option?
Answer this question with a complete sentence.

2. Sherah has just written the music and lyrics for a country song. A major female country music recording star wants to buy the song. She offers Sherah an option: either $5,000 for the rights to both the music and lyrics, or 3% (.03) of sales computed on an average sale price of $5.00 for any recording of the song, whether as a single or as a cut on a CD or tape.

   a. Write 2 equations, one in which Y represents the flat $5,000 payment and the other in which Y equals the amount of money Sherah would receive for X number of sales.

   Equation 1: _____

   Equation 2: _____

   b. Enter these equations at the Y1 = and Y2 = prompt. Ensure you are in the standard viewing screen by pressing **[ZOOM] [6:ZStandard]**. Use the methods developed in the units entitled "Graphing Linear Equations" or "Where Did the Graph Go?" to adjust the viewing rectangle until the point of intersection for the 2 graphs is clearly visible. <u>Describe in detail</u> what you did to obtain a suitable display of the graphs.

   c. Sketch your display and circle the point of intersection. Use the INTERSECT feature of the CALC menu to obtain the X and Y coordinates of the intersection point.

   X = _____   Y = _____

   d. How many recorded copies of the song must be sold for the 2 options to be equivalent? _____

240

3. Trey decides to invest his $5,000 savings for college into two accounts - one of which pays 4% interest annually and the other which pays 10% annually. His first year's interest on his investments is $320.

   a. Using the given information, write 2 equations. Clearly identify what X and Y represent.

   X:

   Y:

   Equation 1: _____

   Equation 2: _____

   b. Enter these equations at the Y1 = and Y2 = prompt. Ensure you are in the standard viewing screen by pressing [ZOOM] [6:ZStandard]. <u>Describe in detail</u> what you did to obtain a suitable display of the graphs and their point of intersection.

   c. Sketch your display and circle the point of intersection. Use the INTERSECT option of the CALC menu to obtain the X and Y coordinates of the intersection point.

   X = _____   Y = _____

   d. How much money was invested at each rate?

4. In 1990 the population of Texas was 16,986,510 people whereas it was 14,225,513 in 1980. It increased by an average of 276,100 people per year in this ten year period. In 1990 the population of Alabama was 4,040,587 people whereas it was 3,849,025 in 1980. It increased by an average of 19,156 people per year in the same 10 year period. Assuming this rate of growth continued, write equations that represent the populations of these states over X number of years (adding the increase for each year to the 1990 population).

Source: The World Almanac and Book of Facts 1995

a. Write 2 equations that relate the above data. Clearly identify what X and Y represent.

X:

Y:

Equation 1: _____

Equation 2: _____

b. Enter these equations at the Y1 = and Y2 = prompt. Ensure you are in the standard viewing screen by pressing [ZOOM] [6:ZStandard]. Describe in detail what you did to obtain a suitable display of the graphs - ZOOMing OUT may not be your best option. Make sure the point of intersection is on the display.

c. Sketch your display and circle the point of intersection. Use the INTERSECT option of the CALC menu to obtain the X and Y coordinates of the intersection point.

X = _____   Y = _____

d. In what year were the populations of the two states the same (round to the nearest year). _____

What was that population (rounded to the nearest whole number).

_____

Solutions: **1a.** Y=50, Y=4X **1c.** X=12.5, Y=50 **1d.** She must visit the pool at least 13 times.
2a. Y=5000, Y=.03(5X) **2c.** X=33333.333 Y=5000 **2d.** 33,334 **3a.** X: 4% account;
Y: 10% account; Eq. 1: X+Y=5000 Eq. 2: .04X+.1Y=320 **3c.** X=3000, Y=2000 **3d.** Three
thousand dollars was invested at 4% and $2,000 was invested at 10%. **4a.** X:number of years;
Y: population; Eq. 1: Y = 16986510 + 276100X; Eq. 2: Y = 4040587 + 19156X
4c. X=-50.38422; Y=3075426.9 **4d.** 1940, 3,075,427

```
┌─────────────────────────────────────────────┐
│                  UNIT #28                     │
│                 FUNCTIONS                     │
└─────────────────────────────────────────────┘
```

In a previous unit, the cost for the first night of the fair was a $5 entrance fee plus $.75 per ride. Translated to an equation, this is $Y = 5 + .75X$ where X is the number of rides and Y is the cost of going to the fair. The cost of attending is dependent on the number of rides, which is an independent choice. Thus Y is our dependent variable and X is our independent variable. We can say that Y is a function of X because for each one value of X that is selected only one Y value is returned. The expression "Y is a function of X" is translated to the symbolic notation $Y = f(X)$.

Domain and Range

1. The X values that can be selected for the above example must all be positive integers since the X variable represents the number of ride tickets purchased. This set of values is called the **DOMAIN** and is written:
    $D = \{X | X \geq 0, X \in \text{Integers}\}$.

EXERCISE SET

Directions: Use the **TRACE** feature to determine the domain for each of the following functions. (Enter all equations A-D at the Y= prompts and turn equations ON and OFF as needed.) TRACE along the path of the graph and examine the X values displayed at the bottom of the screen. These will assist you in determining the domain. You may use any viewing WINDOW you desire, however you may discover that some viewing WINDOWS are more "informative" than others. (Suggestion: view and TRACE on each graph in each of the following WINDOWS - ZStandard, ZDecimal, ZInteger before determining the domain.)

A. $Y = X^2 + 1$ Domain = _____

                        Preferred WINDOW: _____

B. $Y = 3X + 5$ Domain = _____

                        Preferred WINDOW: _____

C. $Y = \sqrt{X + 5}$ Domain = _____

                            Preferred WINDOW: _____

D. $Y = X^3 + 4X^2 + 2$ Domain = _____

                            Preferred WINDOW: _____

Comment on the selection of WINDOW screens. That is, did you find one particular WINDOW better than the others, or did the type of equation dictate your preference of WINDOWS?

3.   As you were TRACING to determine the acceptable X values, the Y values being displayed on the screen were also changing. The Y values, that are dependent on your selection of X (Domain), are called the RANGE.

EXERCISE SET CONTINUED

Directions: Use the TRACE feature and now determine the Range by examining the Y values when you TRACE along the graph. Refer back to problems A - D to determine your preferred WINDOW selection for each equation.

E. $Y = X^2 + 1$ Range = _____

F. $Y = 3X + 5$ Range = _____

G. $Y = \sqrt{X + 5}$ Range = _____

H. $Y = X^3 + 4X^2 + 2$ Range = _____

When an equation is graphed, the VERTICAL LINE TEST can be used to determine if the equation is a function. Remember, to be a function there must be only one X value for each Y value. Thus, when a vertical line is passed across the graph (from left to right) the line will intersect the graph in only one place at a time if the graph represents a function.

4. Enter the equation $Y = X^2 + 2X + 2$ to be graphed and display the graph on the standard viewing WINDOW. To "draw" a vertical line with the calculator, press [2nd] <DRAW>. Press [4:Vertical] (i.e. vertical line). The vertical line is actually displayed on the Y-axis initially. Press the [▸] and [◂] arrow keys to move the vertical line left and/or right across the graph. Since the vertical line does not intersect the graph in more than one place at a time the equation represents a function. Using the [▸] or [◂] keys, return the vertical line to the Y-axis so that it is no longer visible. Each time you graph a new equation you must go back to the DRAW menu to access the Vertical Line option (as well as other options in this menu). Note: Anything that you "DRAW" on your graph via the DRAW menu can only be cleared by pressing [1] for "Clear draw".

EXERCISE SET CONTINUED

Directions: Use the Vertical Line option from the DRAW menu on the graph of each of the following equations to decide if the graph of the equation represents a function. Sketch the graph displayed AND the vertical line at some point on the graph.

I. $Y = -X^2 - 8X - 10$

Function? (yes or no)_____

J. $Y = \frac{1}{2}X^3 + X^2 - 2X + 1$

Function? (yes or no)_____

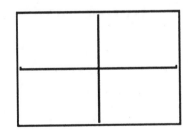

K. $Y = 2\sqrt{X + 4}$

Function? (yes or no)_____

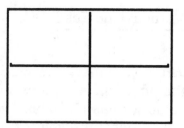

Horizontal Line Test

The HORIZONTAL LINE TEST is used on the graph of a function to determine if the function is one-to-one. A one-to-one (1-1) function has the property that for each Y-value in the range there is one and only one corresponding X-value in the domain. A horizontal line passed across the graph (from top to bottom) will not intersect the graph in more than one place at a time if the graph is that of a one-to-one function.

5. Re-enter $Y = X^2 + 2X + 2$ and display the graph. Press [2nd] <DRAW> [3:Horizontal]. Use the [▲] and [▼] arrow keys to move the horizontal line up and down the graph. (The horizontal line is originally positioned on the X-axis. When you are finished with this option, return the horizontal line to its beginning position on the X-axis.) This is not the graph of a one-to-one function because the horizontal line intersects the graph in <u>two places</u> everywhere except at the vertex.

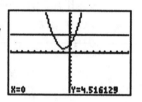

EXERCISE SET CONTINUED

Directions: Use the Horizontal Line option from the **DRAW** menu on the graph of each of the following functions to decide if the graph represents a one-to-one function. Sketch the graph displayed and the horizontal line at a location where it intersects the function in more than one place. If the function is one-to-one you do not need to draw in the horizontal line.

L. $Y = -X^2 - 8X - 10$

1-1 Function? (yes or no)_____

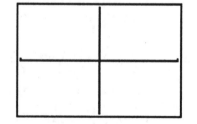

M. $Y = \frac{1}{2}X^3 + X^2 - 2X + 1$

1-1 Function? (yes or no)_____

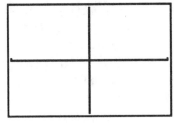

N. $Y = 2\sqrt{X + 4}$

1-1 Function? (yes or no)_____

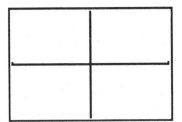

NOTE: If a function is one-to-one, then it has an inverse function that can be determined by interchanging the X and Y variables. The calculator process of graphing an inverse function is outlined in the unit entitled "Exponential Functions and their Inverses". You can refer to items #14-#19 in Unit #31 for specific instructions on drawing inverse graphs. These instructions are independent of the material on exponential functions.

Evaluating Functions

Standard function notation is "f(X)", where f denotes the function, X is the independent variable and f(X) means the function's value at X. If an equation represents the graph of a function, then the Y variable in the equation may be replaced by the f(X) notation.

6. In Exercise A, you should have determined $Y = -X^2 - 8X - 10$ to be a function. We can now write the equation as $f(X) = -X^2 - 8X - 10$. Since the TI-82 will only graph functions, the Y1 is equivalent to the function "f".

7. If we want to evaluate the function $f(X) = -X^2 - 8X - 10$ at X = 2, we write: $f(2) = -(2)^2 - 8(2) - 10 = -30$. Thus when X = 2, f(X) = -30 (i.e. Y = -30). This would yield the ordered pair (2,-30) on our graph of f(X). Verify this by using the TABLES. Be sure your table is set at increments of 1 and enter $-X^2 - 8X - 10$ at the Y1= prompt. Find X = 2 in the table. When X = 2, the table indicates that Y1 = _____.

249

8. The TI-82 uses the equivalent function notation of Y(X) instead of f(X). To evaluate $f(X) = -X^2 - 8X - 10$ at $X = 2$, you must have the f(X) equation entered on the **Y=** screen. (Your equation was entered at the Y1= prompt in #7.) Return to the home screen by pressing **[2nd]** **<QUIT>**. To compute the value for f(X) when $X = 2$, press **[2nd] <Y-vars> [1:function] [1:Y1] [(] [2] [)]** and then **[ENTER]** to compute. The value of the function will be displayed for the X-value of 2. In ordered pair form this would be: (2,-30).

```
Y₁(2)
                      -30
```

EXERCISE SET CONTINUED

Directions: Enter each of the following functions on the "Y=" screen. Evaluate for the indicated value of X using the Y(X) notation. Copy the screen that displays the answer. Record your final information as an ordered pair.

O. Evaluate $Y = 3X^3 - 2X^2 + X - 5$ for $X = -\dfrac{3}{20}$.

Screen display:

Ordered pair:_____ (in decimal form)

Ordered pair:_____ (in fraction form)

P. Evaluate $Y = \sqrt{2X - 5}$ for $X = 3.5$.

Screen display:

Ordered pair:_____

Q. Evaluate $Y = \dfrac{2X + 3}{X^2 + 4X - 5}$ for $X = -\dfrac{1}{2}$

Screen display:

Ordered pair:_____ (in fraction form)

R.   Evaluate Y = $\sqrt{3X + 5}$ for X = -8.

Screen display:

Why did you get this display? Explain carefully.

S.   **Application:** The profit or loss for a publishing company on a textbook supplement can be represented by the function f(X) = 10X - 15000 (X is the number of supplements sold and f(X) is the resulting profit or loss).
i. Use the Y(X) notation and determine the amount of profit (or loss) if 2000 supplements are sold. Copy your screen display to justify your work.

ii. How can you tell if the $5000 is profit or loss?

iii.What would be the profit (or loss) if 1000 supplements are sold? Copy your screen display to justify your work.

9.   Summarize Results: Summarize what you have learned in this unit. Include the following points:
a. how to find the domain and range using the TRACE feature,
a. use of the vertical and horizontal line options and
b. Y(X) notation.

10.	Highlight #4, #5, and #8 for your QUICK REFERENCE list of operations.

<u>Solutions:</u> A. **R** B. **R** C. {X|X≥-5} D. **R** E. {Y|Y≥1} F. **R** G. {Y|Y≥0} H. **R** I. Yes J. Yes K. Yes L. No M. No N. Yes 7. -30 O. (.15, -5.205125), (-3/20, -41641/8000)

P. (3.5, 1.414213562) Q. (-1/2, -8/27) R. $\sqrt{3X + 5}$ is equivalent to $\sqrt{-19}$ when X = -8. The square root function is undefined for negative radicands. S. i. $Y_1(2000)$ 5000 ii. The $5,000 is positive, and therefore a profit. iii. $Y_1(1000)$ -5000 The negative indicates that the $5,000 would be a loss.

There is a *crucial difference* between the TI-82 and the TI-85 when it comes to evaluating functions at a value.

To evaluate a function, say $f(x) = -x^2 - 8x - 10$, you must first enter the function in the [y(x)=] screen. Then you have three options. All take place on the home screen.

Option A: Store the desired value in x then select y1 to be evaluated.

> **Example:** Evaluate $f(x) = -x^2 - 8x - 10$ when x = 2
> **Keystrokes:** First store f(x) as y1, **EXIT** to the home screen, then press
> [2] [STO▶] [x-VAR] [ENTER]   This stores the value of 2 into x.
> [2nd] <VARS> (above [3]) [MORE] [F3](EQU) [▼] (to select y1)
> [ENTER], [ENTER]

Option B: Store the desired value in x then select y1 to be evaluated.

> **Example:** Evaluate $f(x) = -x^2 - 8x - 10$ when x = 2
> **Keystrokes:** First store f(x) as y1, **EXIT** to the home screen, then press
> [2] [STO▶] [x-VAR] [ENTER].   This stores the value of 2 into x.
> [2nd] [ALPHA] <y> [1] [ENTER]

> *****Compare options A and B. They differ only by the way you access y1.

Option C: Instruct the calculator to evaluate *all* the entered functions at a certain value.

> **Example:** Evaluate $f(x) = -x^2 - 8x - 10$ when x = 2
> **Keystrokes:** First store f(x) as y1, [EXIT] to the home screen, then press
> [2nd] <MATH> (above [×]) [F5](MISC) [MORE] [F5](eval)
> then press [2] [ENTER]

The calculator will give you a list of *all* of the currently entered functions evaluated at the value 2. Since f(x) is stored in y1, it will be the first number listed.

UNIT #28 APPENDIX
Casio fx-7700GE

Vertical and Horizontal Line Tests

To draw horizontal or vertical lines using the Casio, see Unit #24 Appendix. However, you cannot *move* these lines up and down, or left and right, as you can with the TI-82.

Evaluating Functions

To evaluate a function, say $f(x) = -x^2 - 8x - 10$, you must first enter the function in the GRAPH screen. Then you store the desired value in x and select y1.

Example: Evaluate $f(x) = -x^2 - 8x - 10$ when x = 2.

Solution: First enter f(x) as y1 in the GRAPH screen. Press [MENU] and select [COMP] and then [AC/ON] to clear the screen. Press [2] [→] [X,θ,T] [ENTER] to store the value of 2 into x. [SHIFT] <VAR> (above [6]) [F1](GRP) [F1](Y). This will place a Y on the screen. Press [1] to use Y1, then [EXE]. The output should be -30.

Vertical and Horizontal Line Tests

While you can draw a vertical line and, in a similar manner, a horizontal line (see
Unit#24 Appendix), you cannot move the line across the graph like in the TI-82. Use
a ruler to apply the vertical and horizontal line tests "by hand."

Evaluating Functions

8. Enter the f(X) equation as a the variable Y using STO. For example, to
    enter f(x) = -X² - 8X -10, press ['] [+/-] [α] <X> [yˣ] [2] [-] [8] [×] [α]
    <X> [-] [1] [0] [ENTER] [STO] [Y]. (Provided Y has not already been
    defined.) To compute the value when X = 2, first define X to be 2
    then in the VAR menu select Y and press [EVAL] to evaluate.

UNIT #29
GRAPHING EQUATIONS WITH RADICALS

This unit will study the square root function by examining the way in which the placement of constants and negative signs affect the graphs as well as the domains and ranges.

1. Enter the equation $Y = \sqrt{X}$ at the Y1 = prompt on your calculator. To get "friendly" numbers for X and Y coordinates, we will use WINDOW values that double those from the ZDecimal screen (ZDecimal x 2):
    XMin = -9.4, XMax = 9.4, XScl = 1,
    YMin = -6.2, YMax = 6.2, YScl = 1.

    Press **[TRACE]** to see your graph. Copy the display.
    Press your right cursor; TRACE and observe the X and Y values at the bottom of the screen.

    What is the domain and the range of the above graph?

    Domain: _____   Range: _____

2. Set your TABLE to begin at -3 and increment by 1. Access the TABLE, and check the Y1 values when X = 0, when X > 0, and when X < 0. Record your observations below:

    X < 0:

    X = 0:

    X > 0:

    For what X-values did you get ERROR messages for Y1? Why?

257

3. Graphs of the square root function will all appear similar, however, the placement of constants affects either/both the domain and/or range. GRAPH and TRACE to determine the domain and range of each equation below. Sketch the graph and record the domain and range in the space provided. (If you have difficulty deciding on the correct domain and range, access the TABLE and check Y1 values as in #2).

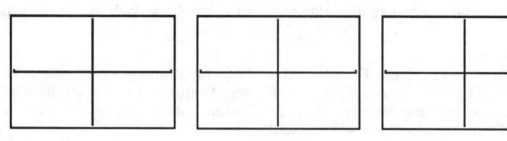

$Y=\sqrt{X+4}$ $Y=\sqrt{X-5}$ $Y=\sqrt{X+3}$

Domain:_____   Domain:_____   Domain:_____

Range: _____   Range: _____   Range: _____

4.

a. Observation: What effect does adding or subtracting a constant, h, to the radicand have on the domain and the range of the equation

$Y = \sqrt{X + h}$?

b. Without graphing, determine the domain and range of the function

$Y=\sqrt{X+\dfrac{1}{2}}$.

Domain:_____

Range: _____

5. GRAPH and TRACE to determine the domain and range of each equation below. Sketch the graph and record the domain and range in the space provided.

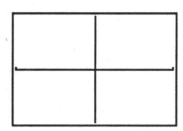   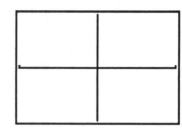

$Y=\sqrt{X}+4$ $Y=\sqrt{X}-5$ $Y=\sqrt{X}+3$

Domain:_____   Domain:_____   Domain:_____

Range: _____   Range: _____   Range: _____

6.   a. Observation: What effect does adding or subtracting a constant, k, to the radical have on the domain and the range of the equation

$Y = \sqrt{X} + k?$

b. Without graphing, determine the domain and range of the function

$Y=\sqrt{X}+\dfrac{3}{2}$

Domain: _____

Range: _____

7.   GRAPH and TRACE to determine the domain and range of each
       equation below. Sketch the graph and record the domain and range in
       the space provided.

$Y = 2\sqrt{X}$ $Y = 6\sqrt{X}$ $Y = .5\sqrt{X}$

Domain:_____   Domain:_____   Domain:_____

Range: _____   Range: _____   Range: _____

8.   Observation: What effect does a positive coefficient "a" in front of the
       radical have on the domain and the range of the equation $Y = a\sqrt{X}$?

9.   What effect do you THINK a negative sign in front of a radical will have
       on the graph of the function? What effect will it have on the domain
       and range?

       Graph $Y = -\sqrt{X}$ and see if your prediction was correct.

Conclusions: Answer each of the following without graphing. You may then go back and check your answers with the calculator.

10.   Write the equation for a square root function with a domain of $\{X|X \geq -8\}$ and with a range of $\{Y|Y \geq 0\}$.

11.   Write the equation for a square root function with a domain of $\{X|X \geq 3\}$ and a range of $\{Y| Y \geq 2\}$.

12.   Write the equation for a square root function with a domain of $\{X|X \leq 0\}$ and a range of $\{Y|Y \geq 0\}$.

13.   Write the equation for a square root function with a domain of $\{X|X \leq 4\}$ and a range of $\{Y|Y \leq 0\}$.

14.   Summarizing Results: Summarize what you have learned about graphing the square root function $Y = a\sqrt{X - h} + k$ from this unit. Your summary should include the following:

      a. The effect on the graph, the domain, and the range of adding or subtracting a constant h to the radicand.
      b. The effect on the graph, the domain, and the range of subtracting X from a constant under the radicand.
      c. The effect on the graph, the domain, and the range of adding or subtracting a constant k from the radical.
      d. The effect on the graph, the domain, and the range of a negative sign in front of the radical.
      e. The effect on the graph, the domain, and the range of a constant factor on the radical.

Solutions: **1.** D: $\{X|X\geq 0\}$; R: $\{Y|Y\geq 0\}$ **2.** ERROR messages when $X<0$; $Y1=0$ when $X=0$; $Y1>0$ when $X>0$. We got ERROR messages when $X<0$ because the square root function is undefined for negative radicands. **3.** D: $\{X|X\geq -4\}$, R: $\{Y|Y\geq 0\}$; D: $\{X|X\geq 5\}$, R: $\{Y|Y\geq 0\}$; D: $\{X|X\geq -3\}$, R: $\{Y|Y\geq 0\}$ **4a.** It affects only the domain; when the constant is added, it shifts the graph that number of units left; when the constant is subtracted, it shifts the graph that number of units to the right. **4b.** D: $\{X|X\geq -1/2\}$, R: $\{Y|Y\geq 0\}$ **5.** D: $\{X|X\geq 0\}$, R: $\{Y|Y\geq 4\}$; D: $\{X|X\geq 0\}$, R: $\{Y|Y\geq -5\}$; D: $\{X|X\geq 0\}$, R: $\{Y|Y\geq 3\}$ **6a.** It effects only the range; adding a constant shifts the graph up that number of units; subtracting a constant moves the graph down that number of units. **6b.** D: $\{X|X\geq 0\}$, R: $\{Y|Y\geq 3/2\}$ **7.** D: $\{X|X\geq 0\}$, R: $\{Y|Y\geq 0\}$; D: $\{X|X\geq 0\}$, R: $\{Y|Y\geq 0\}$; D: $\{X|X\geq 0\}$, R: $\{Y|Y\geq 0\}$ **8.** It affects the how fast the curve rises, which has no effect on either the domain or the range. **10.** $Y=\sqrt{X+8}$ **11.** $Y=\sqrt{X-3}+2$ **12.** $Y=\sqrt{-X}$ **13.** $Y=-\sqrt{4-X}$

262

This unit explores the graphs of quadratic equations, specifically quadratic functions. A quadratic function is an equation in the form $y = ax^2 + bx + c$, where a, b, and c are real numbers. These values (a,b,c) will affect the size and location of the curve. This unit is an exercise in discovery. As you graph each equation, study the size and location of the parabola and compare this information to the coefficients in the equation.

1.   Set your viewing window to ZDecimal x 2 (i.e. [-9.4,9.4] by [-6.2,6.2]) with both scales equal to 1. The TRACE feature will be used to help you discover some of the characteristics of these parabolas. This particular viewing window was selected because the cursor moves will be in tenths of units.

2.   The vertex is the minimum point (or maximum point) on the graph of a parabolic curve. GRAPH and TRACE to find the vertex of each of the parabolas graphed by the given equation. Sketch the graph and record the coordinates of the vertex as ordered pairs.

Y = X²   Y = 3X²   Y = (¼)X²

vertex:_____   vertex:_____   vertex:_____

3.   Observation: What effect does the coefficient on the X² term have on the graph of the equation?

263

4.   GRAPH and TRACE to find the vertex on each of the equations. Sketch the graph and record the coordinates of the vertex.

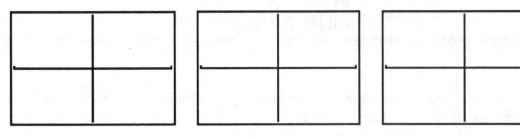

$Y = X^2 + 3$

vertex:_____

$Y = X^2 + 1$

vertex:_____

$Y = X^2 - 2$

vertex:_____

5.   Observation: What effect does the constant term have on the graph of the equation?

6.   GRAPH and TRACE to find the vertex on each of the equations. Sketch the graph and record the coordinates of the vertex.

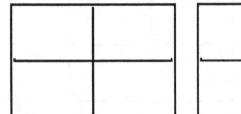

$Y = (X + 6)^2$

vertex:_____

$Y = (X + 1)^2$

vertex:_____

$Y = (X - 3)^2$

vertex:_____

7.   Observation: When a value is added or subtracted to the X <u>before</u> the quantity is squared, what effect does it have on the graph?

8.   GRAPH and TRACE to find the vertex on each of the equations. Sketch the graph and record the coordinates of the vertex.

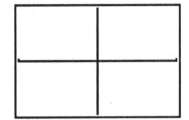 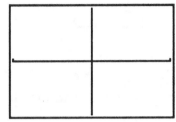 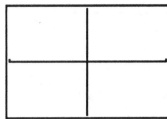

$Y = (X + 3)^2 + 2$ $Y = (X + 3)^2 - 2$ $Y = (X - 3)^2 + 2$

vertex:_____   vertex:_____   vertex:_____

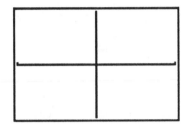 $Y = (X - 3)^2 - 2$

vertex:_____

9.   Based on your observations from the previous problems, what should the vertex of the parabola $Y = (X + 115)^2 - 38$ be? **DO NOT ATTEMPT TO ANSWER THIS QUESTION BY GRAPHING THE EQUATION!** You are to use the information gathered in the previous problems to determine the vertex.

_____

10.   Compare each pair of equations by graphing on the same graph screen.

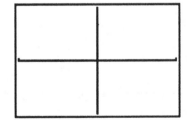   a. $Y = 3(X + 2)^2 + 2$

b. $Y = (X + 2)^2 + 2$

What effect did the "3" have on the shape of the graph?

265

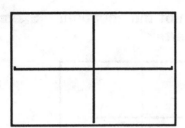

a. $Y = (¼)(X + 2)^2 + 2$

b. $Y = (X + 2)^2 + 2$

What effect did the "¼" have on the shape of the graph?

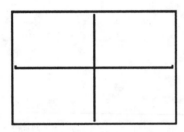

a. $Y = 3(X + 2)^2 + 2$

b. $Y = -3(X + 2)^2 + 2$

What effect did the negative sign on the 3 have on the shape/orientation of the graph?

CONCLUSIONS: Answer the following questions without graphing the equation on the graphing calculator. You may then go back and check your answers with the calculator. Some questions may have more than one correct response.

_____11. Which of these graphs will have its vertex at the origin?
        a. $Y = (X - 5)^2$
        b. $Y = X^2$
        c. $Y = (4/5)X^2$
        d. $Y = 2X^2 + 7$
        e. $Y = 4(X + 3)^2 - 7$

_____12. Which of these is the graph of $Y = X^2$ translated (shifted) two units to the left of the Y-axis?
        a. $Y = 2X^2$
        b. $Y = (X + 2)^2$
        c. $Y = X^2 - 2$
        d. $Y = (X + 2)^2 - 2$
        e. $Y = (X - 2)^2$

_____13. Which of these is the graph of $Y = X^2$ translated 2 units down from the X-axis?
  a. $Y = 2X^2$
  b. $Y = (X + 2)^2$
  c. $Y = X^2 - 2$
  d. $Y = (X + 2)^2 - 2$
  e. $Y = (X - 2)^2$

_____14. Which of these graphs has a maximum point?
  a. $Y = 2X^2$
  b. $Y = -2(X - 2)^2$
  c. $Y = X^2 - 2$
  d. $Y = (X + 2)^2 - 2$
  e. $Y = 2 - X^2$

15.
  a. Based on your observations, for the parabola $y = a(x - h)^2 + k$, what effect does $a > 0$ and $a < 0$ have on the orientation of the parabola?

  b. What effect does the size of $|a|$ have on the size of the parabola?

  c. The value of h will shift (translate) the parabola which direction?

  d. The value of k will shift (translate) the parabola which direction?

Solutions: 2. (0,0), (0,0), (0,0) 3. It affects the width of the parabola. 4. (0,3), (0,1), (0,-2) 5. It moves the parabola up or down the Y-axis. 6. (-6,0), (-1,0), (3,0) 7. It shifts the parabola left or right along the X-axis. 8. (-3,2), (-3,-2), (3,2), (3,-2) 9. (-115,-38) 10. The "3" made the graph "skinnier". The "¼" made the graph "fatter". The negative sign turned the graph "upside down". 11. b,c 12. b,d 13. c,d 14. b,e 15a. If $a > 0$, the parabola opens up and has a minimum. If $a < 0$, the parabola opens down and has a maximum. 15b. $|a|$ determines the width of the parabola. 15c. "h" shifts the parabola right or left. 15d. "k" shifts the parabola up or down.

This unit will examine the graphs of exponential functions and their inverses, logarithmic functions. All graphs should be displayed on the ZDecimal x 2 screen.

1. An exponential function is a function of the form $Y = f(X) = b^X$ where b is a positive real number not equal to 1.

2. We will first examine the function when b (the base) is larger than 1, $b > 1$. GRAPH and TRACE each function below. Sketch the display carefully.

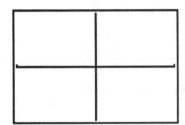   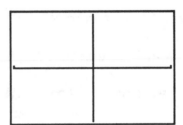   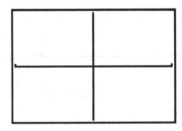

$Y = 2^X$ $Y = 3^X$ $Y = 8^X$

3. Access your TABLE (increment by 1). Scroll up and down while examining the values for Y.

    a. As X increases, what happens to Y?

    b. As X decreases, what happens to Y?

    c. Will the value of Y ever be equal to 0? Why or why not?

    d. State the domain and the range of the functions.

       Domain:_____   Range:_____

    e. How are the above 3 graphs

       similar:

       different:

4.   We will now examine the function when b (the base) is larger than 0 but less than 1, $0 < b < 1$. GRAPH and TRACE each function below. Sketch the display carefully.

$Y = (1/10)^X$ $Y = (1/2)^X$ $Y = (4/5)^X$

5.   Access your TABLE (ensure that it is set to increment by 1). Scroll up and down while examining the values for Y.

   a. As X increases, what happens to Y?

   b. As X decreases, what happens to Y?

   c. Will the value of Y ever be equal to 0? Why or why not?

   d. State the domain and the range of the functions.

      Domain:_____   Range:_____

   e. How are the above 3 graphs

      similar:

      different:

6.   The exponential functions we have graphed are continuous. Moreover, they are either continuously increasing or continuously decreasing. Explain what the word "continuous" means in the above contexts, i.e. define continuous, continuously increasing function, and continuously decreasing function.

7. All of the graphs above have the same Y-intercept. What is it?
_____

8. Why is the point with coordinates (0,1) on each of the above graphs?

9. If you multiply an exponential function by a constant, what do you THINK will happen to the graph?

10. Graph $Y1 = 2^X$ and $Y2 = 6(2^X)$ on your calculator and sketch the display. Compare Y2 to Y1.

   a. Were the domain or range affected?_____

   b. Did the Y-intercept change?_____

   c. What is the Y-intercept of Y2?_____

11. Predict the Y-intercept for $Y = 8(2^X)$._____

12. What if a constant were added to the function $Y = 2^X$? Describe the manner in which the constant will shift the graph.

   positive constant:

   negative constant:

13. Graph the exponential function $Y = 2^X + 3$ on your calculator and sketch the display below. Were your predictions from #12 correct?

If a function is one-to-one, i.e. passes both the vertical and horizontal line tests outlined previously in the "Functions" unit, it will have an inverse function.

14. Consider the function Y = 2X + 3. Enter 2X + 3 at the Y1 = prompt on your calculator (using the ZDecimal x 2 screen). Copy your display.

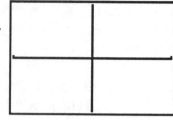

This is a one-to-one function because the graph passes both the vertical and horizontal line tests.

15. Algebraically find the inverse below by (a) interchanging the X and Y variables and (b) solving for Y.

16. Your inverse equation should be $Y = \frac{1}{2}X - \frac{3}{2}$.

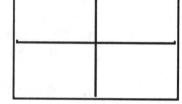

Graph this equation at the Y2 = prompt.
Copy the display of the two graphs.

17. At the Y3 = prompt, enter X to graph Y = X. Your display should look like the one displayed at the right.

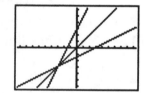

Label the lines as Y1, Y2, and Y3.

Y1 and Y2 are symmetric across the line Y = X (our Y3 line). This will always be true of a function and its inverse.

18. To use the calculator to **DRAW** the inverse function, you must have your function entered on the **Y =** screen. **Since your function is entered at Y1, go back and delete Y2 and Y3 before proceeding.** Press [2nd] <QUIT> to return to the home screen.

19. Instruct the calculator to **DRAW** the inverse of Y1 (DrawInv): Press [2nd] <DRAW> [8:DrawInv] [2nd] <Y-vars> [1:Function] [1:Y1] [ENTER]. Y1 will be graphed first and the INVERSE of Y1 will be drawn second. The inverse that is drawn is the same line as the one you graphed at Y2, however, because it is drawn (and not graphed from the Y = screen) you will not be able to interact with the graph. That is to say, you will not be able to TRACE, use INTERSECT, ROOT, VALUE, TABLE, etc.

The inverse of an exponential function is a logarithmic function. To find the inverse of the exponential function ($Y = b^X$) we will interchange X and Y. This yields the equation $X = b^Y$ ($b > 0$, $b \neq 1$). Presently we do not have an algebraic method for solving this equation for Y. We will use the **DrawInv** option from the DRAW menu to draw the inverses of these functions. These inverse functions are logarithmic functions, defined as $Y = \log_b X$.

20. Graph $Y = 2^X$. Following the instructions in #18 and #19, draw the inverse of Y1. Sketch your final display of both functions and label Y1 and INV Y1 on the graph.

Y1: Domain:_____
Range:_____

INV Y1: Domain:_____ Range:_____

21. Why are the domain and range in the INV Y1 graph the reverse of the domain and range of the Y1 graph?

Recall that the Y1 graph had a Y-intercept of 1 and no X-intercept. Although you cannot TRACE on INV Y1, what appear to be the X and Y intercepts for the **inverse of Y1**?

22. Graph $Y = 0.5^X$. Draw the inverse of Y1. Sketch your final display of both functions and label Y1 and INV Y1 on the graph.

Y1: Domain:_____ Range:_____

Y-intercept:_____

INV Y1: Domain:_____ Range:_____

X-intercept:_____

273

23. Summarizing Results: Summarize what you have learned in this unit. Your summary should:
a. state the definition of an exponential function,
b. discuss the graphs of exponential functions when b > 0 (your discussion should address domain, range and Y-intercept)
c. discuss the graphs of exponential functions when 0 < b < 1 (your discussion should address domain, range and Y-intercept)
d. discuss why the case of b = 1 is excluded in the definition of an exponential function
e. state the definition of a logarithmic function (i.e. the inverse of an exponential function)
f. discuss the relationship between the domain, range and intercepts of an exponential function and its inverse.

23. Highlight #18 and #19 in this unit for your QUICK REFERENCE list of operations.

Solutions:

2.

3a. As X increases, Y increases. **3b.** As X decreases, Y decreases.

3c. No: a non-zero base raised to any power is a non-zero number. **3d.** D:$\mathbb{R}$, R: $\{Y|Y>0\}$
3e. Answers may vary.

4.

5a. As X increases, Y decreases. **5b.** As X decreases, Y increases.

5c. No: a non-zero base raised to any power is a non-zero number. **5d.** D:$\mathbb{R}$, R: $\{Y|Y>0\}$
5e. Answers may vary. **6.** A continuous function is one in which there are no gaps (or holes). A continuously increasing and/or decreasing functions have no turns (i.e. no relative maximums or minimums). **7.** 1 **8.** When the exponent is zero the value of the exponential expression is one.
9. Answers may vary. **10a.** No **10b.** Yes **10c.** 6 **11.** 8 **12.** positive constant: shifts graph up; negative constant: shifts graph down

13.   **20.** Y1: see 3d.; INV Y1: domain: $\{X|X>0\}$, range: $\mathbb{R}$ **21.** Because the X and Y variables were interchanged. The X-intercept = 1 and there is no Y-intercept.

22. Y1: same as 5d., Y-intercept = 1; INV Y1: domain: $\{X|X>0\}$, range: $\mathbb{R}$, X-intercept = 1

275

To draw inverses using the TI-85, first enter the desired function as y1. Press [EXIT] until the GRAPH menu clears from the screen. Then, from the GRAPH menu ([GRAPH] [MORE]), select [DRAW] [MORE] [MORE] [F2] (for [DrInv]). This will place the command **DrInv** on the home screen. Place y1 next to it by: [2nd] [ALPHA] <y> [1] , press [ENTER].

The Casio does not have the ability to *draw* inverses in the same way as the TI-82. However, by using the parametric mode of the calculator, we can accomplish the same thing. Every function has a parametric representation, and you will learn about this process in a future math class. For now, you will be using just enough of the idea to graph functions and their inverses.

Example: Graph $y = 2^x$, its inverse, and the line $y = x$

Procedure: Get into Parametric mode: [MENU] [COMP] [EXE] [SHIFT] [SET UP] [F3] (for [PRM]) [EXIT]

Set the appropriate range: Note that the screen must be "square." [RANGE] then [F1] (for [INIT]) to get

[EXIT]

Graph $y = 2^x$: [GRAPH] [X,θ,T] [SHIFT] < , > (above [→]) [2] [^] [X,θ,T] [)] [EXE]

Switch back to the COMP screen: [G↔T]

Graph the inverse: [GRAPH] [2] [^] [X,θ,T] [SHIFT] < , > (above [→]) [X,θ,T] [)] [EXE]

Switch back to the COMP screen: [G↔T]

Graph the line $y = x$:   [GRAPH] [X,θ,T] [SHIFT] < , > (above [→]) [X,θ,T] [)] [EXE]

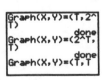 

277

This process can be used to graph the inverse of *any* function. To graph *f(x)* and its inverse, simply change the variable to **T**, and enter

$$\textbf{Graph(X,Y)} = \textbf{(T, } \textit{f}\textbf{(T))}$$
$$\textbf{Graph(X,Y)} = \textbf{(}\textit{f}\textbf{(T),T)}$$
$$\text{and} \quad \textbf{Graph(X,Y)} = \textbf{(T,T)}.$$

This will show the graph, the inverse, and the line $y = x$, respectively. It is a bit cumbersome, but after some practice, it is actually rather easy. Unfortunately, you will not be able to TRACE any of these graphs.

Recall that the HP 48G does not have a TABLE feature. Use TRACE with the graph set at ZINTG to find the values.

INVERSES

The HP 48G does not have a DRAWINV feature. But we can use the HP to symbolically find the inverse of a function and then graph this function. First enter the function you wish to invert, with the X and Y interchanged, as a variable. You can name this variable F1. Press [→] [SYMBOLIC] [▼] [▼] [▼] (OK) to open the ISOLATE A VARIABLE application. At the EXPR line, press (CHOOS) and then select F1. At the VAR line, enter Y. Be sure the RESULT option is set to Symbolic. Check the PRINCIPAL option. Then press (OK). Store the result as a variable. You can call this variable F2. Then use the PLOT application to graph the inverse and function.

For example, suppose you want to graph $Y = \dfrac{1}{2}X - \dfrac{3}{2}$ and it inverse. First save the

function as the variable F. Then save $X = \dfrac{1}{2}Y - \dfrac{3}{2}$ as the

variable F1. Enter the ISOLATE A VARIABLE application and set the screen so it looks like the one at the right.

Press (OK). The inverse function now appears on level 1 of the stack. Press [α] <F> [2] [STO]. Enter the PLOT application and CHOOS the variables F and F2 at the EQ line. Your PLOT screen should look like the one at the right.

EXPONENTIAL INVERSES

Use the same procedure as above to graph an exponential function and its inverse. Recall that to change the contents of a variable that you enter the new expression on level 1 of the stack then press left-shift key and the menu key corresponding to the variable.

I.   Match the graph with the appropriate equation. Each graph in this section is graphed in the standard viewing window.

A 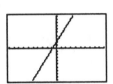   **B** 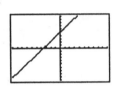   **C** 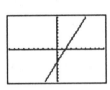   **D**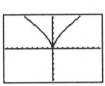

E 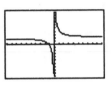   **F** 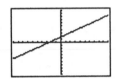   **G** 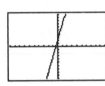   **H**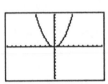

I 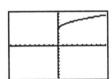   **J** 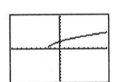   **K**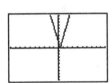

___   1. $Y = \dfrac{7}{3X} + 2$ ___   6. $Y = 2X + \dfrac{5}{3}$

___   2. $Y = 2X^{\frac{2}{3}}$ ___   7. $Y = \sqrt{2X + 5}$

___   3. $Y = \dfrac{2X + 5}{3}$ ___   8. $Y = \dfrac{2X^2}{3}$

___   4. $Y = \sqrt{2X} + 5$ ___   9. $Y = \dfrac{7}{3}X - 2$

___   5. $Y = |-2 + 5X|$ ___   10. $Y = \sqrt{2}X + 5$

II.   Match each equation to its appropriate graph. <u>Viewing rectangles</u> <u>may</u> <u>vary</u>. The X and Y scales are indicated if they are not equal to 1.

_____   1. $Y = |X + 2|$

_____   2. $Y = X^4 - 6X^2 + 9$

_____   3. $Y = -\sqrt{2 - X}$

_____   4. $Y = X^2 + 2$

_____   5. $Y = |X| + 2$

_____   6. $Y = \sqrt{X} + 2$

_____   7. $Y = X^3$

_____   8. $Y = \sqrt{X + 2}$

_____   9. $Y = (X + 2)^2$

_____   10. $Y = -X^3 - 1$

A

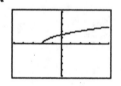

B

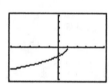

C

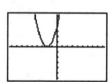

X & Y scale = 2

D

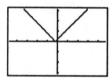

X & Y scale = 10

E

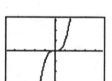

F

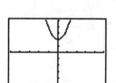

G

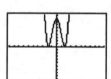

H

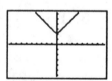

I

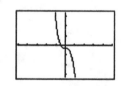

Xscl = 2, Yscl = 3

J

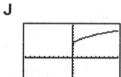

Xscl = ½, Yscl = ½

K

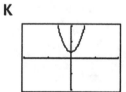

X & Y scale = 5

282

Solutions: I. 1. E 2. D 3. F 4. I 5. K 6. A 7. J 8. H 9. C 10. B
II. 1. D 2. G 3. B 4. F 5. H 6. J 7. E 8. A 9. C 10. I

CORRELATION CHART

UNITS	PRE-REQ. UNIT(S)	CORRELATING CONCEPT
#33 Statistics: Plotting Paired Data	#1	Paired data and the types of statistical plots available for display are explored.
#34 Frequency Distributions	#1, #33	Frequency distributions and the use of the calculator to construct histograms
#35 Line of Best Fit	#11, #34	A calculator approach to graphing linear regression equations and interpreting displayed data.

INTRODUCTION OF KEYS

Unit Title	Keys	
33: Statistics: Plotting Paired Data	STAT	1:Edit
		4:ClrList
	STAT PLOT	1:Plot 1
	ZOOM	9:ZoomStat
34: Frequency Distributions	STAT ▸ CALC	3:SetUp
		1:1-Var Stats
35: Line of Best Fit	STAT ▸ CALC	5:LinReg(ax + b)
	VARS	5:Statistics
		EQ
		7:RegEQ

UNIT #33
STATISTICS: PLOTTING PAIRED DATA

Descriptive statistics is the collection, presentation and summarization of data. This unit examines paired data information and the type of plots available on the TI-82 for displaying this data.

1.   The STAT feature can be used as a tool for graphing ordered pairs of numbers. Unlike the graph screen, or TABLE feature, STATPLOT will allow you to plot individual ordered pairs of numbers. The graph screen and TABLE feature are dependent on functions being entered on the **Y=** screen.
NOTE: Before proceeding, delete all entries on the **Y=** screen.

2.   The ordered pairs (2,3), (-5,7), (4,-2), (0,-4) and (5,8) can be arranged in a table of values:

X	Y
2	3
-5	7
4	-2
0	-4
5	8

We will create two lists in the STAT menu to represent each list in our table of values.

3.   Begin by pressing **[STAT]** and **[1:Edit]**. Use the ► arrow key to cursor over to the sixth list (L6) to observe the content of each of the lists. You must clear the data from the lists before beginning any problem.

4.   For demonstration purposes, suppose that the first and second lists (L1 and L2) have data stored in them. To clear these lists, press **[2nd] <QUIT>** to return to the home screen, press **[STAT] [4:ClrList]** and press **[2nd] <L1> [,] [2nd] <L2> [ENTER]**.

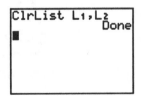

Another approach is to press **[STAT] [1:Edit]**, use the ▲ key to highlight L1 and press **[CLEAR]**. Cursor down below L1 to see the list clear.

5. The data (ordered pairs in your table of values) can be entered in the lists on the Edit screen in two ways. We will examine the first approach using the data listed in #2.

a. Press [STAT] [1:Edit]. Enter each X value, in order, in the L1 list. (Press [2] [ENTER] [(-)] [5] [ENTER] [4] [ENTER] [0] [ENTER] [5] [ENTER].)

b. Use the ▶ arrow key to cursor over to the L2 list column and enter the Y values. Double check your paired data to be sure that they are the same as your original table of values.

c. Press [2nd] <STATPLOT> [1:Plot 1]. Press [ENTER] to turn on the first plot. Use the ▼ arrow key to highlight the first icon after **Type** (press [ENTER]), the ▼ arrow key to highlight the first entry (L1) after **Xlist** (press [ENTER]), the ▼ and ▶ arrow keys to highlight the second entry (L2) after **Ylist** (press [ENTER]), and finally, the ▼ arrow key to highlight the first entry after **Mark** (press [ENTER]). Plotted data points can be represented by ☐, +, or a •. We selected ☐ for easy visibility.

d. These ordered pairs can be plotted using either of the options 4, 6 or 9 on the **ZOOM** menu, or by entering values on the **WINDOW** screen. You must be sure that the X values in your list are between the Xmin and Xmax of your WINDOW and that the Y values in your list are between the Ymin and Ymax of your WINDOW.

e. Press [ZOOM] [6:ZStandard] to view the ordered pairs in the standard viewing WINDOW.

f. Press [TRACE]; the screen at the right is displayed. P1 is displayed in the upper right corner of your screen, indicating that Plot 1 is turned on. The ◀ and ▶ arrow keys allow you to move from point to point, displaying the ordered pair values at the bottom of the screen. This type of display is called a Scatter Plot.

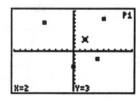

g. Press [ZOOM] [9:ZoomStat] and the points are replotted with the calculator automatically resetting the viewing WINDOW to include all the data points.

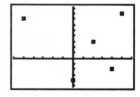

6. Clear lists L1 and L2 using one of the methods in #4.

7. We will now enter a new table of values using a different approach.

    a. The table of values is   X   Y . To enter the entire set of values

X	Y
3	6
-2	1
4	-2

at once, return to the home screen. Use the **{ }** to list all the X values and **STO**re them in L1.

To accomplish this, press [2nd] < { > [3] [,] [(-)] [2] [,] [4] [2nd] < } > [STO▸] [2nd] <L1> [ENTER].

    b. Store the list of Y values in L2. Your screen should now look like the one at the right.

```
Plot1
 ■ Off
Type: ▦ ⚬ ⬝⬝ ⬛
Xlist: ▉L2 L3 L4 L5 L6
Ylist: L1 ▉L3 L4 L5 L6
Mark: ▣ ♦ ·
```

    c. Check **Plot 1** under the **STATPLOT** menu to be sure all items are selected as in #5c.

    d. Press [ZOOM] [9:ZoomStat]. Sketch the graph displayed.

NOTE: Remember to clear the lists in L1 and L2 before beginning a new problem. (See #4 for directions.)

8. Enter the table of values by pressing the **STAT** key and display the data points using the **ZoomStat** viewing WINDOW. Sketch the graph displayed.

X	Y
1	3
2	-5
0	0
-1	-2
-3	2
5	1

9. **Application:** On the first five days of January, the daily highs were 22°, 19°, 12°, 17° and 13°. Let days 1 through 5 be your X values and the temperatures be the Y values.

    a. Display the data points using the **ZoomStat** Window. Sketch the display.

b. Press [2nd] <STATPLOT> [1:Plot 1] and move the cursor to the second **Type** of display and press [ENTER] to select. Press [GRAPH]. As you can see, this mode of display connects the data points in sequence. This **XYline** graph allows you to clearly see the temperature pattern over the 5 day period. Sketch this display.

10. Using {-3, -2, -1, 0, 1, 2, 3} for the list of X values, create a table of values for the equation Y = 3X - 2. Enter the information in the lists on the **STAT** menu. HINT: Enter the list of X values under **L1** and 5X - 2 at the Y1= prompt. Now type Y1(L1) on the home screen and press [ENTER]. Press [STAT] [1:Edit] to view your lists.

    a. Display the data points as a Scatter Plot graphed in the **ZoomStat** viewing WINDOW and sketch your screen display.

    b. Now display this same data, connecting the data points with a straight line. This is called the **XYline**. To do this, return to the **STATPLOT** menu by pressing [2nd] <STATPLOT> and [1] to select Plot1. Plot1 is already turned ON, cursor down to **TYPE** and right once, to the second icon. Press [ENTER] to select the icon (which represents the **XYline**). Press [GRAPH] and sketch your screen display.

11. Summarize Results: Summarize what you have learned in this unit. Include the following:
    a. how to clear lists in **STAT**
    b. how to enter data in the lists
    c. how to select your Plot display
    d. explanation of the difference between Scatter Plots and XYlines.

12.   Highlight #4, #5a and #7a for your QUICK REFERENCE list of operations.

Solutions: **7d.**   **8.**   **9a.**   **9b.**   **10a.**

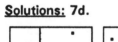   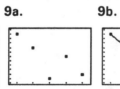

10b.

The Statistics capabilities of the TI-85 are different from those of the TI-82. Consult the TI-85 GUIDEBOOK for a complete discussion of its capabilities. This appendix will explain how to do *some* of the same topics discussed in this unit.

The **STAT** menu (press [STAT]) on the TI-85 has five options. First select [F2](EDIT). Press [ENTER] twice to accept the names **xStat** and **yStat**. (The TI-85 can accept a large number of lists, more than the 6 allowed by the TI-82. Each must be given a different name at the initial step. The default names are **xStat** and **yStat**, as above. If you use more than one list, each list must have a different name.)

Enter the data as an ordered pair, (x, y). The screen at the right shows the first two pairs of values entered.

To make a scatterplot of this data, first sort the data in order of the x values by pressing [F3](SORTX), then press [2nd] <M3>(DRAW). Select [F2](SCAT) and you will see:

Notice that the points that are plotted are just single-pixel dots. Unfortunately, the TI-85 does not have the capability to display data points as +'s or other symbols. Further, you cannot TRACE these data points. Neither is there a ZOOMSTAT function on the TI-85. Simply look through your lists and make sure that the x values and y values are contained between the xMin, xMax, yMin, and yMax values of your RANGE settings.

To display the **xyLINE**, you simply select [F3](xyLINE) from the DRAW menu under the STAT feature.

UNIT #33 APPENDIX
Casio fx-7700GE

The Statistics capabilities of the Casio are different from those of the TI-82. Consult the _fx-7700GE_ Owner's Manual for a complete discussion of its capabilities. This appendix will explain how to do *some* of the same topics as discussed in this unit.

From the MAIN MENU, select **REG**. The screen should then show: It is important that the calculator be set for **S-data: STO** and **S-graph: DRAW**. If not, press [SHIFT] <SET UP> and make the appropriate changes.

Now enter the paired data from the table in #2 of Unit #33 as follows:

1.   Press [AC/ON] to clear the screen.

2.   Press [Range] and adjust the range values to contain all the data points in question. There is no ZOOM STAT feature to assist in this process, so you must enter the correct Minimum and Maximum values to see the data.

3.   Press [EXIT] to return to the main screen.

4.   Type [2] [F3](,) [3] [F1](DT)
               [(-)] [5] [F3](,) [7] [F1](DT)
               [4] [F3](,) [(-)] [2] [F1](DT)
               [0] [F3](,) [(-)] [4] [F1](DT)
               [5] [F3](,) [8] [F1](DT)

         The idea in the above is to enter the data, pair by pair, and pressing [DT] to enter the data point. As each point is entered, it is plotted on the axes. The Casio does not have the capability to use +'s or other symbols to represent the data points. You also cannot TRACE the points and see their coordinates.

In the **REG** mode of the Casio, you cannot draw the **xyLINE** as on the TI-82.

The STAT application is used to store statistical data. Press [→] [STAT] to enter the STAT application and select the Single-variable mode. The data is stored in a variable named ΣDAT. At ΣDAT press the EDIT menu key. Then enter a table of values. Use the cursor keys to move from entry to entry and press [ENTER] to enter the data. For example, the data for #2 would look like this:

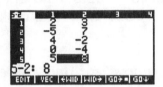

Press [ENTER] to exit back to the SINGLE-VARIABLE STATISTICS window. Press (OK) to store the information. If you want to delete the data, press [DEL] (OK) while the data is highlighted.

To plot the data, enter the PLOT application from the HOME window by pressing [→] [PLOT]. Change the TYPE to SCATTER by selecting (CHOOS) then moving down until SCATTER is highlighted and press [ENTER]. Your plot screen should now look like this:

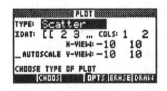

Press (DRAW) to plot the data. Note that (0, -4) does not appear on the plot because the vertical axis is "hiding" it. If you want to see it, just uncheck AXES in the PLOT OPTIONS window. You can also enter data in the PLOT application.

The HP 48G *does not* have a XYLine option that will connect the data points in sequence.

This unit will continue to look at ways to present and summarize data. The last unit examined scatter plots and XYlines as means of presenting two variable statistics. One variable data can be presented as a histogram. A histogram is a vertical bar graph with no spaces between the bars. The horizontal axis (X-axis) displays the values (L1) and the vertical axis (Y-axis) displays the frequencies (L2) with which the L1 values occur.

CONSTRUCTING HISTOGRAMS

The following steps should be followed when displaying one variable statistics as a histogram. Read through these steps: **DO NOT** implement the steps until you have been given a problem with data.

1. Construct a frequency table of the data:

2. Delete all entries on the **Y=** screen and clear all lists L1 through L6 by pressing **[STAT] [4:ClrList] [2nd] <L1> [,] [2nd] <L2>** etc. To determine which lists, if any, have values entered, press **[STAT] [1:Edit]** to view the lists.

3. Turn on **STATPLOT** and set up for a histogram. Press **[2nd] <STATPLOT> [1:Plot 1]** (**ON** should be highlighted), cursor down to **Type** and over to highlight the fourth icon, press **[ENTER]**, cursor down and press **[ENTER]** to highlight L1 for **Xlist** and L2 for **Frequency**. Your screen display should correspond to the one at the right.

4. Set the calculator for one variable statistics by pressing **[STAT] [▶]** (for CALC menu) **[3:SetUp]**. L1 should be highlighted for the **Xlist** and L2 for the **Frequency**. Use the cursor arrows to move the cursor over the desired entry and press **[ENTER]** to highlight.

5. Enter the data from your frequency table into the calculator's STAT list display by pressing **[STAT] [1:Edit]** and entering the values in the L1 column and their respective frequencies in the L2 column.

6.   Set the WINDOW: Be sure that your values fall <u>between</u> the Xmin and Xmax, that the Ymin = 0 and Ymax is greater than your largest frequency value. The Xscl will determine how many values are grouped in each vertical bar. The calculator <u>requires</u> that $\dfrac{Xmax-Xmin}{Xscl} \le 47$. Set the Xscl = 1 unless otherwise noted. Yscl will equal to 1 since it represents frequencies. You may TRACE on the histogram; note that the number "n" will indicate how many values are in the bar that your trace cursor is located on.

7.   Translate the calculator's histogram to a hand drawn and <u>labeled</u> histogram.

EXAMPLE 1

Suppose a 10 point quiz in an intermediate algebra class yielded the following scores:

10,9,6,7,8,8,9,8,7,7,6,7,8,8,8,7,9

Construct a histogram to represent the frequency distribution. **(The numbers of each step correspond to each of the step numbers listed previously.)**

1.   Begin by constructing a frequency table:

X	10	9	8	7	6
frequency	1	3	6	5	2

2.   Clear all lists L1 through L6.

3.   Turn on **STATPLOT 1** and set up for a histogram display.

4.   Set the calculator for one variable statistics.

5.   Enter the data from your frequency table into the calculator's STAT list display by pressing **[STAT] [1:Edit]** and entering the grades in the L1 column and their respective frequencies in the L2 column.

6.   Set your WINDOW values as follows: Xmin = 5, Xmax = 11, Xscl = 1, Ymin = 0, Ymax = 7, Yscl = 1. Your screen should look like the one displayed at the right.

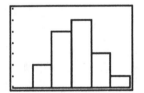

7.   If we were to draw this histogram by hand it would look like the one at the right:

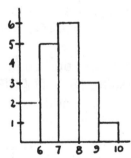

296

Some of the statistical features about these grades that would be of interest are displayed on the **1-Var Stats** option found by pressing **[STAT] [▶]** (for CALC menu) and **[1:1-Var Stats] [ENTER]**.

$\bar{x} \approx 7.8$ ($\bar{x}$ is the arithmetic average or mean)

median = 8 (Med is number that divides ranked data into two equal groups. You may cursor down to display it.)

$\sigma x \approx 1.05$ (σx = standard deviation: measures the dispersion of the data about the mean)

EXAMPLE 2

The scores below are the final course scores for an intermediate algebra class:

> 83, 78, 82, 91, 84, 75, 84, 96, 80, 68, 79, 78, 85, 89, 95, 82, 88, 74, 76, 73, 74, 64, 66, 93, 69, 87

A frequency distribution is not appropriate here because of the wide variance of the scores. However, a histogram representing the number of A's, B's, etc. would be informative. Enter the data only in L1.

NOTE: Data can be sorted in ascending order by pressing [STAT] [2:Sort A()] [2nd] <L1> [ENTER].
When you have frequencies listed in L2, you would need to enter Sort A(L1,L2) so that the frequencies are sorted with their respective values.

If you do not use a frequency distribution then you must be sure that the frequency value is set to "1" under **STATPLOT 1** and **STAT/CALC SetUp**. (Do this now.)

The local college uses a 10 point grading scale (i.e. A: 90-100, B: 80-89, etc.) and the public schools use an 8 point grading scale (i.e. A: 93-100, B: 85-92, C: 77-84, D: 69-76, F: below 69). We will construct two histograms: one to represent the distribution of letter grades for a 10 point scale and the other to represent the distribution of letter grades for an 8 point scale.

a. The Xscl will determine which values (grades) are grouped in each bar. For your first histogram, set the WINDOW values as follows: Xmin = 60, Xmax = 110, Xscl = 10, Ymin = 0, Ymax = 10, Yscl = 1. Ymax may need to be increased if any one grade group has more than 10 scores in it. Press **[GRAPH]**. Sketch the histogram displayed for this WINDOW. Use TRACE to determine the number of people that made an A?_____, a B?_____, a C?_____, a D? _____, failed?_____

b. For your second histogram, set the WINDOW values as follows: Xmin: 60, Xmax: 110, Xscl: **8**, Ymin: 0, Ymax: 10, Yscl: 1. Ymax may need to be increased if any one grade group has more than 10 scores in it. Sketch the histogram displayed for this WINDOW.

How many people made an A?_____
a B?_____, a C?_____, a D? _____, failed?_____

EXERCISE SET

Directions: For each problem complete the following:
a. organize the data in a frequency table,
b. enter the data in the calculator using the STAT feature,
c. sort the data in ascending order
d. copy the display of the histogram (Set your own WINDOW values; record the values used. Be sure to set both the Xscl and Yscl at 1.)
e. construct a hand drawn and labeled histogram, and
f. record the indicated information.

A. A freshmen girl's P.E. class is surveyed. The following shoe sizes were recorded:
8,7,8,8,7,7,7,6,5,6,6,6,7,9,5,7,10,9,9,8,7,7,7,6,6,8,6,7,6

X							
frequency							

Calculator display:

WINDOW :

```
MINIMA FORMAT
 Xmin=
 Xmax=
 Xscl=
 Ymin=
 Ymax=
 Yscl=
```

Hand Drawn (and labeled) Sketch:

$\bar{x} \approx$ _____ (nearest whole size)

$\sigma x \approx$ _____ (nearest hundredth)

median = _____

B. The daily highs for the month of February were recorded as:

68, 68, 73, 72, 73, 73, 68, 68, 73, 69, 69, 73, 73, 72, 73, 73, 68, 68, 69, 69, 69, 70, 69, 73, 70, 70, 70, 69, 71

Calculator display:

WINDOW:

Hand Drawn (and labeled) Sketch:

What is the average temperature? _____ (to the nearest tenth of a degree)

What is the median temperature?_____

Solutions:

Example 2a.   **Example 2b.**   **Exercise A.**

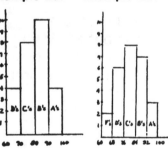

$\bar{x} \sim 7$
$\sigma x \sim 1.22$
median = 7

Exercise B.

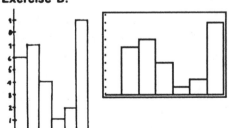

average temp. 70.4
median temp. 70

299

CONSTRUCTING HISTOGRAMS

To use the HISTOGRAM option on the TI-85, press **[STAT] [F2](edit) [ENTER] [ENTER]** and enter each value into the x-list, along with its associated frequency in the y-list. Then press **[2nd] <M3>(DRAW)** and select **[F1](HIST)** from under the DRAW menu.

EXAMPLE 1

The histogram on the TI-85 appears below. The range settings appear to its left. Notice that space was left at the bottom of the screen to allow for the menu bars.

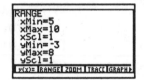 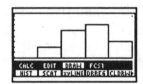

To calculate the one-variable statistics, select **[CALC]** from the **[STAT]** menu and press **[ENTER]** twice, the press **[F1](1-VAR)**. You can find the mean and the standard deviation from this, but the TI-85 does not determine the median of the data.

EXAMPLE 2

You can, similar to the TI-82, change the width of the bars of the histogram by changing the value of xScl in the RANGE window. The TI-85 does not allow you to trace this histogram, as does the TI-82. To determine the frequency of each score, you can approximate the height of the histogram bar by reading off tic marks on the y-axis.

To use the HISTOGRAM option on the Casio, select **SD** from the MAIN MENU. Press **[AC/ON]** to clear the screen. The Casio is rather sensitive in the order in which you must proceed. With a little practice, you will become quite comfortable with this procedure, but if your histogram bars do not appear as they should, refer back to this procedure. We will proceed step-by-step.

1. *CLEAR DATA*

> If the data screen is blank, proceed to step 2, otherwise clear out the old data as follows: **[F2](EDIT) [F3](ERS) [F1](YES)**

2. *SET UP RANGE*

> Press **[Range]** and enter the appropriate values for the data you are working with. A few notes:

>> (a) Xscl determines the column width and must agree with the width of the columns for the data you are using.
>> (b) Each column graphs to the *right* of its associated x-value. Therefore, the Xmax value you enter must be one full column width beyond your last x-value.
>> (c) If you plan to trace your data, it helps if the Ymin value is a negative value. Then, when the x and y values are displayed at the bottom of the screen, they do not cover the histogram bars.

> Make a note of the **Xmin, Xmax,** and **Xscl** values. You will need them later.

> Press **[EXIT]**.

3. *SET UP AND RECALIBRATE CALCULATOR MEMORY*

> First clear the old statistics settings:
> **[Shift] <CLR> [F2](Scl) [EXE] [F6](CAL)**
> (This clears the old memory and recalibrates the calculator)
> Now set up the new statistics settings:

> Calculate $n = \dfrac{Xmax - Xmin}{Xscl}$

Define the correct number of *possible* columns:
   [Shift] <Defm> *n* [EXE]
(This tells the calculator to allocate one memory location for each possible column, whether or not you have data for each one.)

4. *ENTER DATA*

Refer to EXAMPLE 1 from Unit #34. Follow the instructions from step 6 and set up the Range, referring above to *SET UP RANGE.* Then enter the data from step 1 as follows:

   [1] [0] [F1](DT)
   [9] [F3](;) [3] [F1](DT)
   [8] [F3](;) [6] [F1](DT)
   [7] [F3](;) [5] [F1](DT)
   [6] [F3](;) [2] [F1](DT)

Notice that the default frequency is 1, and for higher frequencies, you simply enter, for example, 9;3 then press [DT]. This enters 9 with a frequency of 3.

5. *DRAW THE HISTOGRAM*

Press [Graph] [EXE] to see the histogram.

(This display uses the following RANGE: Xmin = 5, Xmax = 11, Xscl = 1, Ymin = -1, Ymax = 10, Yscl = 1.)

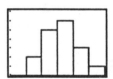

6. *TRACE THE HISTOGRAM*

Press [F1](Trace) to trace the height of each histogram bar. This can only be done immediately after graphing the histogram.

Note: The Casio does not combine data into columns in the same way as the TI-82. It is best to combine the data into the appropriate frequency distribution, then enter it into the Casio.

To calculate the one-variable statistics, select [F4](DEV) from the **SD** menu, then select the mean, standard deviation, or median using the menu keys.

UNIT #34 APPENDIX
HP 48G

CONSTRUCTING HISTOGRAMS

As in the previous unit, enter the STAT application and select the Single-variable
mode. Again, the data is stored in the variable named ΣDAT. Delete the previous
data. To construct a histogram, you must enter each data into one column.
Therefore, the number of entries is equal to the number of data. Unlike the TI-82,
you cannot make a list of values with a list of frequencies.

To draw a histogram, enter the PLOT application and change the TYPE to Histogram.
In the PLOT OPTIONS screen, make sure that H-TICK and V-TICK is set to the
distance you want between tick marks and that PIXELS is unchecked. In the PLOT
screen, set WID to the width you want, and set H-VIEW and V-VIEW to appropriate
values.

EXAMPLE 1

The data in EXAMPLE 1 would appear like this:

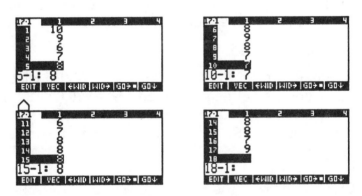

As you can see, there are seventeen entries in column 1.

To draw a histogram, enter the PLOT application and change the TYPE to Histogram.
In the PLOT OPTIONS screen, make sure that H-TICK and V-TICK is set to 1 and
that PIXELS is unchecked. In the PLOT screen, set WID to 1, the H-VIEW from 5 to
11, and V-VIEW from 0 to 7. Then press **(DRAW)**.

Your screen should look like this:

Some of statistical features about these grades are available in the SINGLE-VARIABLE STATISTICS window. Set the TYPE to Population. Check the statistical features that you are interested in.

EXAMPLE 2

Enter the PLOT application of TYPE Histogram. Delete the data in ΣDAT and enter the final scores into the first column.

a.   Set WID to 10, H-VIEW from 60 to 110, V-VIEW from 0 to 10. Under PLOT OPTIONS set H-TICK to 10, V-TICK to 1, and uncheck PIXELS. Then press (DRAW) to draw the histogram.

To find the number of people that made a certain grade, TRACE does not work very well on the HP 48G. Instead, open the STAT application and select the Frequencies option. ΣDAT should already be the data you entered earlier. Set X-MIN to 60 since our minimum value is 60. Set BIN COUNT to 5 since there are 5 groups of 10 from 60 to 110. Set BIN WIDTH to 10 since we are grouping every 10 points starting from 60. Press (OK). This returns you to the stack. There are now two new arrays on level 1 and level 2. The entries of the array on level 2 give the number of points that fell in each bin (in this case, each group of 10 starting from 60) from lowest to highest. To view the array, press [▲] twice to move up to the second level then press (VIEW) to view the array. The first entry gives the number of data points that fell in the 60 to 69 range, the second entry gives the number of data points that fell in the 70 to 79 range, and so on.

b.   For the second histogram, enter the PLOT application of TYPE Histogram. Use the same ΣDAT that you had already entered. Set WID to 8, H-VIEW from 60 to 108, V-VIEW from 0 to 10. Under PLOT OPTIONS set H-TICK to 8, V-TICK to 1, and uncheck PIXELS. Then press (DRAW) to draw the histogram.

Enter the FREQUENCIES window in the STAT application. Set BIN COUNT to 6 since there are 6 groups of 8 from 60 to 108. ΣDAT should already be the data you entered earlier. Set X-MIN to 60 since our minimum value is 60. Set BIN WIDTH to 8 since we are grouping every 8 points starting from 60. Press (OK).

```
┌─────────────────────────────────────────────┐
│                  UNIT #35                     │
│              LINE OF BEST FIT                 │
└─────────────────────────────────────────────┘
```

In this unit we will determine the relationship between two variable quantities (data points) in different types of problems. If the data points lie in a straight line then the equation of the line passing through these points can be used to predict unknown data points. We want to consider situations where the relationship is <u>not</u> linear. Data points in this type of problem produce a scatter plot. We will examine a line (regression equation) that comes "close" to passing through all the data points. The most frequently used algebraic approach to determine regression equations is the principle of least squares. We will use the calculator instead.

PROCEDURE FOR REGRESSION EQUATION

Listed are the steps necessary to determine a regression equation with the calculator. (You may find many textbooks that refer to this line as the "line of best fit".)

1. Clear all data from lists L1 - L6 and delete all entries from the **Y=** screen.
2. Enter paired data in lists L1 and L2.
3. Press [STAT] [▶] (to highlight CALC) [3:Setup]. Under 2-Var Stats highlight L1 for the Xlist and L2 for the Ylist.
4. Turn **ON STATPLOT 1**, highlight the first icon under **TYPE** (scatter plot), L1 for Xlist and L2 for Ylist. There are three choices for the type of Mark, we have chosen to use □, the first icon, for visual clarity.
5. Press [ZOOM] [9:ZoomStat] to view your scatter plot. If you have a mechanical pencil, use a piece of the lead - or a similarly short piece of spaghetti, to approximate the "line of best fit". The next steps will instruct the calculator to compute the equation of this "line of best fit" and graph it.
6. Press [STAT] [▶] [5:LinReg(ax + b)] [ENTER]. On the home screen the equation y = ax + b, along with the corresponding variable values, will be displayed. This equation needs to be entered on the **Y=** screen. You do not need to try to remember these long variable values, as the calculator can be directed to copy this equation to the **Y=** screen. Directions for this are in the next step.
7. Press [Y=] (You must be at the Y= screen to copy the equation. Clear all Y= entries now.) [VARS] [5:Statistics] [▶] [▶] (to highlight EQ for equation) [7:regEQ].
8. Press [GRAPH] to display the scatter plot and the "line of best fit". The viewing WINDOW was automatically set to ZoomStat in #5. As long as the **STATPLOT** is turned **ON** you will be able to TRACE on the scattered data points and on the line using the ▲ or ▼ keys to move from data point to line. You will not, however, be able to TRACE beyond the ZoomStat viewing window unless you edit the WINDOW values. For this reason, we will examine information relating to the line by using the TABLE feature.

```
```

The following table gives the winning time (in seconds) for the Men's 400-Meter Freestyle swimming event in the Olympics from 1904 to 1992. The Olympics were not held during the years 1916, 1940, and 1944 due to the two World Wars. A linear regression equation will be used to predict the winning times for each of these years.

Step numbers correspond to the numbers listed above.

YEAR	1904	1908	1912	1920	1924	1928	1932	1936	1948
TIME	376.2	336.8	324.4	326.8	304.2	301.6	288.4	284.5	281

1952	1956	1960	1964	1968	1972	1976	1980	1984	1988	1992
270.7	267.3	258.3	252.2	249	240.3	231.9	231.3	231.2	227	225

(Source: The World Almanac and Book of Facts 1995)

1. Clear data from lists.

2. Enter years in L1 list and winning time in the L2 list.

3. Go to **STAT/CALC Setup** and highlight L1 for Xlist and L2 for Ylist.

4. Turn on **STATPlot 1** (be sure all others are OFF). Your STATPLOT screen should look like the one displayed.

5. Press **[ZOOM] [9:ZoomStat]** to view the scatter plot. Your display should correspond to the one at the right. Place your pencil lead on the screen to approximate your line of best fit. Draw in this line on the screen at the right.

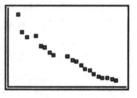

6. Determine the linear regression equation (the equation for your line of best fit). The following information should be displayed on your home screen.

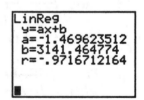

7. Copy this equation to the **Y=** screen.

8. Press **[GRAPH]** to view the scatter plot and the "line of best fit". See screen display at the right.

To predict the winning times in the missing years, set the table to start at 1916 (increment by 4 - Olympics occur every 4 years). Press **<TABLE>**, scroll to view the predicted winning times for the missing Olympic years. The winning time in 1916 would have been _____, in

1940:_____ and in 1944:_____. This regression line comes "close" to passing through the set of points. Therefore, your predictions are approximations at best.

Directions: For each problem, follow the steps on the first page of this unit. Sketch the graph screen that displays the scatter plot and the calculator's line of best fit. Use the TABLE to answer the question(s) posed.

A.   The chart below gives the year and winning time (in seconds) for the Men's 1000-Meter Speed Skating event. Predict the winning time for the 1998 Winter Olympics.

YEAR	1976	1980	1984	1988	1992	1994
TIME	79.32	75.18	75.8	73.03	74.85	72.43

(Source: The World Almanac and Book of Facts 1995)

Graph display:

Winning time prediction for the 1998
Winter Olympics:_____

B.   A survey was taken of notably tall buildings on the east coast, mid-west, and west coast. The survey compares height, in feet from sidewalk to roof, to number of stories (beginning at street level). If the Empire State building has 102 stories, use the given information to predict the height of the building.
(Source: The World Almanac and Book of Facts 1995, actual height of the Empire State building is 1250 ft.)

BUILDING	STORIES	HEIGHT (FT.)
Baltimore U.S. Fidelity and Guaranty Company	40	529
Maryland National Bank (Baltimore, MD)	34	509
Sears Tower (Chicago, IL)	110	1454
John Hancock Building (Chicago, IL)	100	1127
Transamerica Pyramid (San Francisco, CA)	48	853
Bank of America (San Francisco, CA)	52	778

Graph display:

Predicted height of the Empire State
Building:_____

C.   The statistics in the table below represent dollars per 100 pounds for
     cattle. Predict the price per 100 lb. of cattle for the years 1995 -
     2000.

YEAR	1940	1950	1960	1970	1975	1979	1980	1984
PRICE	7.56	23.30	20.40	27.10	32.20	66.10	62.40	57.30

1985	1986	1987	1988	1989	1990	1991	1992	1993
53.70	52.60	61.10	66.60	69.50	74.60	72.70	71.30	72.60

Source: The World Almanac and Book of Facts 1995)

Graph display:

Predicted price per 100 lb.:

1995:_____   1998:_____

1996:_____   1999:_____

1997:_____   2000:_____

Solutions: Example: 1916: 325.67, 1940: 290.4, 1944: 284.52
A. Winning time: 71.499

B. height: 1281.1

C.

1995: 73.48   1998: 77.38
1996: 74.78   1999: 78.67
1997: 76.08   2000: 79.97

To determine the regression line, select **[STAT] [F1]** (for **[CALC]**) **[ENTER] [ENTER]** then **[F2](LINR)**.

To plot the regression line, first make a scatterplot of the data, then press **[GRAPH]** and select the **[F1](y(x)=)** menu. Clear all functions, then next to **y1=** press **[2nd] <VARS> [MORE] [MORE] [F3](STAT)**. Select **RegEq** and press **[ENTER]**. Then graph the function. It will appear along with the data points.

Plotting the Regression Line

To plot a regression line on the Casio, you first enter your data as described in the appendix to Unit #33. Then press [Graph] [SHIFT] [F4](Line) [1] [EXE]. To redraw this graph, press [Cls] [EXE] then [SHIFT] <CLR> [F2](Scl) [EXE] [F6](CAL). This will re-plot the points. Press [Graph] [SHIFT] [F4](Line) [1] [EXE] again to graph the regression line again.

To find the values of a and b in the regression equation $y = ax + b$, press [F6](REG) and choose [F1] or [F2].

PROCEDURE FOR REGRESSION EQUATION

1. Enter STATISTICS application by pressing [→] [STAT]. Press [▾] [▾] (OK) to select the Fit Data option.

2. In the FIT DATA window, delete the data in ΣDAT then press (ENTER) and enter the paired data in column one and two as you did in Unit#33. X-COL should be set to 1 and Y-COL should be set to 2. At the MODEL entry, the selection, Linear Fit, should already be there. If not, press (CHOOS) and select it.

3. Press (OK). You will return to the stack. The right-hand side of the equation of best fit appears on level 3 of the stack.

4. To graph the equation, enter the PLOT application and set the TYPE to Scatter. Press (DRAW).

5. The viewing range is probably not set correctly so press (ZOOM) [NXT] (to scroll the menu list) (ZAUTO). ZAUTO will set the viewing range so all the data will be displayed.

6. Press [→] <MENU> to display the PICTURE menu. Press (STATL) to draw the line of best fit. The TRACE feature can be used on the line.

EXAMPLE

Follow the above steps to find and plot the equation of best fit.

3. You should get this screen after step 3:

6. You should get this screen after step 6:

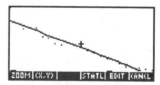

To predict the winning times in the missing years, use the right hand side of the equation of best fit which should be in level 3 of the stack. Press [▲] three times to get to level 3. Press (PICK) to move the equation to level 1. Press [ON] to get out of the stack. You can store the expression as a variable, say F. Then evaluate it at X = 1916, X = 1940, and X = 1944 as you did in Unit#8.

TROUBLE SHOOTING

```
ERR:SYNTAX
1:Goto
2:Quit
```

Your directions to the calculator were incorrect. Look for misplaced parentheses, use of subtract sign (blue key) for the negative sign (gray key), etc.

* * * * *

```
ERR:DIVIDE BY 0
1:Goto
2:Quit
```

You have entered an expression whose denominator is zero. Division by zero is undefined.

* * * * *

```
ERR:WINDOW RANGE
1:Quit
```

The values entered in the WINDOW screen are inappropriate. Be sure that
a. Xmin < Xmax and Ymin < Ymax and
b. you have used the gray negative key for negative values and not the blue subtract key.

* * * * *

```
ERR:DOMAIN
1:Goto
2:Quit
```

The value that has been selected for X is not acceptable (i.e. not in the domain). Example: $\sqrt{-4}$

* * * * *

```
ERR:BOUND
1:Quit
```

a. The LOWER bound you set is not less than (i.e. to the left of) the UPPER bound on the graph screen. Remember, lower bound refers to an X value that is less than the point you are determining.
b. When establishing upper and lower bounds, your "guess" was not selected <u>between</u> the established bounds.

* * * * *

```
ERR:SIGN CHNG
1:Quit
```

a. No real root - graph does not intersect the X-axis between the lower and upper bounds you established.
b. The two graphs do not intersect.
c. The two graphs do intersect but the intersection is not visible on the display screen.

* * * * *

```
ERR:BAD GUESS
1:Quit
```

When using the CALC menu, the "guess" you entered was not acceptable. Try again, this time entering your guess as close as possible to the desired point.

* * * * *

```
ERR:STAT PLOT
1:Quit
```

a. Data entered incorrectly
b. Turn STATPLOT OFF if graphing functions.
c. STAT SetUp ([STAT] [▸] (CALC) [3:SetUp]) not compatible with data lists and/or STATPLOTS.

* * * * *

```
ERR:STAT
1:Quit
```

$\dfrac{Xmax - Xmin}{Xscl}$ <u>must be less than or equal to</u> 47

TO THE OWNER OF THIS BOOK:

We hope that you have found *Explorations in Beginning & Intermediate Algebra using the TI-82*, useful. So that this book can be improved in a future edition, would you take the time to complete this sheet and return it? Thank you.

School and address: _____

Department: _____

Instructor's name: _____

1. What I like most about this workbook is: _____

_____

_____

2. What I like least about this workbook is: _____

_____

_____

3. My general reaction to this workbook is: _____

_____

4. The name of the course in which I used this workbook is: _____

_____

5. The name of the main textbook I used in the course is: _____

6. Were all of the units of the workbook assigned for you to read? _____

   If not, which ones weren't? _____

7. In the space below, or on a separate sheet of paper, please write specific suggestions for improving this workbook and anything else you'd care to share about your experience in using the workbook.

_____

_____

_____

_____

_____

Optional:

Your name: _____ Date: _____

May Brooks/Cole quote you, either in promotion for *Explorations in Beginning & Intermediate Algebra using the TI-82* or in future publishing ventures?

Yes: _____ No: _____

Sincerely,

Deborah J. Cochener
Bonnie M. Hodge

FOLD HERE

- -

BUSINESS REPLY MAIL

FIRST CLASS   PERMIT NO. 358   PACIFIC GROVE, CA

POSTAGE WILL BE PAID BY ADDRESSEE

ATT: *Deborah J. Cochener and Bonnie M. Hodge*

Brooks/Cole Publishing Company
511 Forest Lodge Road
Pacific Grove, California 93950-9968

FOLD HERE